**Brushless Motors and Controllers**

Die Deutsche Nationalbibliothek verzeichnet diese Publikation in der Deutschen Nationalbibliografie; detaillierte bibliografische Daten sind im Internet über dnb.d-nb.de abrufbar.

Impressum:

Roland Büchi, 2012

Herstellung und Verlag: Books on Demand GmbH, Norderstedt

ISBN: 978-3-8448-0107-1

German edition:
'Brushless-Motoren und -Regler', publisher: vth, Verlag für Technik und Handwerk

**Preamble**

Brushless DC motors and controllers have begun an unparalleled triumph in recent years in all technical fields and also in model construction. It is not just that they are now an alternative to the brushed or DC motors. No, they have in many fields almost completely driven these out of business.

But what are the apparently strong arguments that speak for this new technology? As a model constructor one simply buys a finished motor and controller unit for the airplane, car or boat model. And then one enjoys, together with the LiPo battery, a huge difference when comparing the available power and the flight or drive time with what was previously possible. The only thing you can see from the outside as the difference is that between the controller and the brushless motor there are now three wires instead of only two, as was the case with brushed motors.

This book is intended to show the interested reader how a brushless motor works, what its properties are, and what the reasons are that have led to its success. The basic principle is discussed first, before all the key terms such as kV and rpm/V, operating voltage, load and no-load current, torque, turns, electrical and mechanical power, losses, efficiency, etc. are explained.

A brushless motor can't work without a brushless controller. The batteries supply only a DC voltage, whereas a brushless motor requires a so-called three-phase AC voltage. Thus it is microprocessor technology together with powerful transistors which enables the operation.

To increase speed properly, the controller must have information on the rotor position, the so-called phase zero crossing. The measurement is done in most cases without sensors, directly via the motor windings. In many applications there are, however, Hall sensors which perform this measurement. They have some advantages in the startup phase, but the disadvantage of the higher price. Both types will be taken into account in the book.

Since industrial drives are also often designed as brushless motors with one of these two measurement methods, the book is also of interest to technicians.

However, the theory here is to be executed only as far as it helps the reader to understand the principles. For better understanding, practical examples will always be presented. At the end of most chapters, each of the key findings is presented in short form. In this way, also the technical layman gains a basic understanding of this technology.

Of course, there are also practical examples of motor and controller combinations discussed in the book. Several cases of controllers and motors of different power are examined. The load cases of the different types of models – propellers on model aircraft, helicopters in Governor mode, ship propellers of model boats or the acceleration and braking of cars – play a big role. They all find their place here, and the specialties of these drive units are also considered here.

There are several ways in which the settings of brushless controllers can be carried out. Some work on an integrated programming mode, which communicates with the user via switches, LEDs and beeps. Often, parameters can be saved using the throttle stick of the remote control. More elaborate versions make use of so-called programming boxes, which are plugged directly to the controller for setting the parameters. Another type of programming is done via USB port on the PC. With the included software it is possible to change a variety of parameters graphically on the screen.

Also important are practical tips for proper installation in the model and tips on wiring. Since in operation both the motor and the controller generate power loss and therefore heat, the discussion about the proper cooling of both units is very important. Notes for the purchasing decision and the right procedure for troubleshoot are also essential and complete the book. References are listed at the end

# Contents

**1. Introduction**     **6**
- 1.1 Electro-mechanical energy converter ........................... 6
- 1.2 Differences between brushed and brushless motor ...... 8
- 1.3 No brushless motor without brushless DC controller .. 10
- 1.4 Brushless DC motor or just brushless motor? ............. 12

**2. Working principle of the brushless motor**     **13**
- 2.1 Active principle ................................................................ 13
- 2.2 Action of force on a current-carrying conductor in a magnetic field .................................................................. 19
- 2.3 Force action in practice on the example of an inrunner ........................................................................... 22
- 2.4 More stator slots and magnetic poles on the examples of outrunners 6S8P and 12S14P .................................. 27
- 2.5 Star connection, delta connection ................................. 37
- 2.6 Findings in brief ............................................................. 45

**3. Characteristics of the brushless motor**     **47**
- 3.1 kV, rpm/V ....................................................................... 47
- 3.2 Winding resistor ............................................................. 52
- 3.3 Motor characteristics ..................................................... 57
- 3.4 Turns .............................................................................. 66
- 3.5 Motor power .................................................................. 66
- 3.6 Braking operation .......................................................... 71
- 3.7 Efficiency ....................................................................... 73
- 3.8 Findings in brief ............................................................. 79

**4. Brushless DC Controllers**     **81**
- 4.1 Power stage ................................................................... 82
- 4.2 Timing and phase measurement .................................. 89
- 4.3 Maximum current, current and temperature measurement ................................................................. 97
- 4.4 BEC voltage ................................................................. 100
- 4.5 Low voltage measurement .......................................... 104
- 4.6 Operation modes, stick position = voltage or speed .. 105
- 4.7 Microcontroller and setting the parameters ............... 105
- 4.8 Findings in brief ........................................................... 109

## 5. Practical examples — 112
- 5.1 Practical tips for cabling .......................................... 112
- 5.2 Graupner Compact drive .......................................... 114
- 5.3 Modification of a monster truck to a brushless motor .......................................... 120

## 6. Sources of error — 126
- 6.1 Short to frame, short-circuited coil, short circuit ........ 126
- 6.2 Bearing and shaft .......................................... 128
- 6.3 Defective power stage of the controller .................... 129

# 1. Introduction

Brushless DC motors – or BLDC motors in short – are an indispensable part of modern drive technology. Other names are also EC (electronic commutated) motor or synchronous motor. They have for many years been standard for actuating drives or also for electrical power generation. However, the use of miniature motors up to the kilowatt range was reserved for a long time for brushed (DC) motors. At the beginning of the technology of power electronics, DC was much easier to control. But as considered in more detail later, brushless motors need multi-phase, mostly three-phase AC. It was at that time very complex to produce that in the different frequencies and voltages required. So their use was only worthwhile for larger power classes.

But today's microprocessor technology combined with efficient transistors of the smallest size allowed since the beginning of the 21st Century that ever-smaller brushless motors have been competitive in price compared to DC motors. As will be shown, they are superior to DC motors in many ways, either because of the low wear or the high efficiency. Anyone who has ever replaced a DC drive with a brushless drive will therefore never look back on the other technology.

This is particularly true for model construction. In the last years of the 20th Century almost all small electric motors were brushed. Here, the change to brushless drives took place within a few years. The many benefits of this technology ensured that they became not only the standard drives for boat and car models, but even for flight models. This is particularly interesting because there they need to compete with combustion engines. For a long time these seemed to be undisputed in terms of power density.

## 1.1 Electro-mechanical energy converter

Electric motors are so-called electro-mechanical energy converters. They convert electrical energy or power into mechanical energy or power. Both electrical and mechanical power are important for the user. Electrical power is relevant for the calculation of how long it takes until the battery is empty. The mechanical power is what drives the model; that is, what one can see or feel. Figure 1 shows the relations. As can be seen from this, energy conversion cannot be done in practice without loss. There are always losses in the motor.

It can be deduced that for each electrical motor the following equation is valid:

$P_{EL} = P_{MECH} + P_L$

*Figure 1: Power distribution*

$P_L$ is the total power loss. As will be shown later, it consists of the losses in the winding, the losses in the iron and the friction between the shaft and bearings.

The mechanical power in the motor operation is always smaller than the electrical power. The manufacturers are very interested in maintaining the ratio between the mechanical and electrical power as high as possible. This is the motor efficiency. It's not just about the optimum use of electrical power. If high efficiency is achieved, then this also means that the power loss during operation is small. Then the heat in the drive is low. This increases the life span. Due to the special technology of brushless motors, the efficiencies are in most cases in the operating range between 80% and 90%, and often even more. In comparison to that, brushed motors or DC motors achieve efficiencies which lie between 50% and 80%. Brushed motors with comparable power thus have much higher power dissipation. They will thus in operation under the same conditions get warmer or have to be built bigger and heavier to dissipate the heat.

**Power classes**
Later it will be shown that the stator surface, i.e. the non-rotating part, determines how big the maximum power loss of the motor may be. Thus, if the efficiency is high, with a given stator surface a large mechanical power can also be achieved. Of course most of the motors used in models lie in the power class of a few watts to several hundred watts. But there are also drives with up to 15 kW and more built and operated successfully. Together with the high-performance accumulators and controllers of today the largest model aircraft, boats and cars can also be operated. The brushless drive is the drive of the present and future. With the development of even stronger powered systems, the number of combustion engine powered models will decrease further.

## 1.2    Differences between brushed and brushless motor

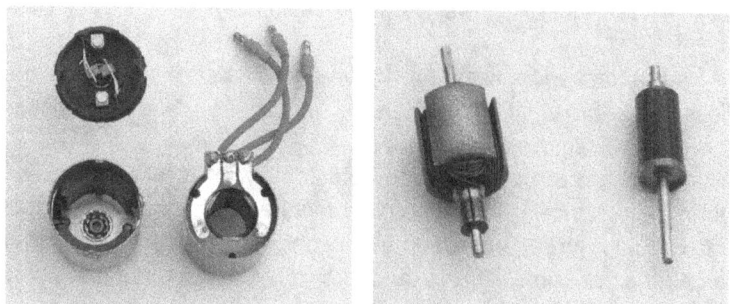

Figure 2: Stators and rotors of brushed and brushless motors

Figure 2 shows stators and rotors of brushed and brushless motors. On the left of the left-hand image is a stator and brushes of a brushed motor, while to the right is the stator of a brushless motor. The right image shows on the left a rotor of a brushed and on the right a rotor of a brushless motor. Both motor types are so-called inrunners. The fixed part, the stator, is located outside and the rotating part, the rotor, is inside.

Both motors are moving with the same principle of Lenz's law, as will be discussed in detail in Section 2.2. Under this principle, (permanent) magnets and current-carrying conductors always exert forces on each other. With a brushed motor, the permanent magnets are always mounted on the stator. The current-carrying conductors are always located on the turning part. However, the magnetic field caused by the permanent magnet and the current through the conductor only exerts a force in the direction of rotor rotation when they are in the correct orientation relative to one another. This means that the current will be reversed during the rotation. This is in technical language also called 'commutation'. The brushes are needed to establish a contact between the stator and the wires of the turning rotor. In addition, they commutate the current. They are usually made of graphite. To minimize wear and enhance contact, they are often enriched with copper, silver or other metals. Brushed motors have the advantage that they can be driven with a DC voltage, as provided by a battery or an accumulator. The commutation with the brushes then ensures the correct flow direction. Therefore, these motors are also called direct current (DC) motors.

But it is precisely this connection between the rotor and the stator which is the weak point of these motors. On the one hand, the power transfer requires a certain pressure between brushes and rotor, which increases the friction and reduces the efficiency, while on the other hand the lower efficiency then means that the motor generates more power loss in operation. The resulting heat just wears off again the main cause of low efficiency, namely these same brushes. Inductances in the motor also ensure that the abrupt change of current flow during commutation causes large overvoltage peaks. The resulting so-called brush fire disturbs the receiver of the remote control. The brushed motor must therefore be dejammed.

With brushless motors, the permanent magnets are always attached to the rotating part, the rotor. The current-carrying conductors are however always located on the stator. This situation is therefore exactly the reverse to the one with brushed motors. Placed in this way, the current does not need to be transferred to the rotor. The brushes are therefore eliminated.

One might almost ask at this point why the technology has made such a long detour via the brushed motor. Ultimately, it's much easier if the wearing parts of the brushes are not needed at all. The answer lies in the commutation. A brushless motor can't operate directly with a DC voltage from a battery. The commutation must be resolved in a different way. What is done with brushed motors mechanically is done with brushless motors through the electronics of the brushless DC controller. Commutate current also means changing current direction, or indeed using alternating current, AC. The brushless controller turns the DC battery power to AC. Only this ensures that the motor is turning.

It is a fact that it can be seen very often in modern technology: a problem which was earlier solved purely mechanically is now replaced by a combination of electronics and much simpler mechanics. Other examples are the wheel hub motor in bicycles, instead of a chain drive and a gear, or 'fly by wire' of a large aircraft, instead of mechanical transmission rods. An example in model construction is the replacement of the complicated helicopter mechanism by four speed-controlled propellers of a quadrocopter. The brushless motor is thus in good and modern company.

## 1.3 No brushless motor without brushless DC controller

The above title is immediately clear. A brushless motor can't yet rotate on its own because the highly important commutation is absent. This is immediately clear to anyone who holds such a motor with three wires in his hand the first time. How can he connect these with only the two wires of the battery? Therefore a regulator is mandatory. These two things are inextricably linked.

In the early days of model construction with brushed motors and remote controls with few channels, it was still possible (but not advisable) to connect the motor directly to the battery, without the possibility of intervention. For the above reasons, today's brushless drives must always consist of motor and controller. Figure 3 shows such a standard combination; Figure 4 is a schematic representation of this.

As shown in Chapter 4, the brushless DC controllers almost always have a so-called BEC (Battery Eliminator Circuit) connector. This provides the receiver with energy and can be seen in the figures too.

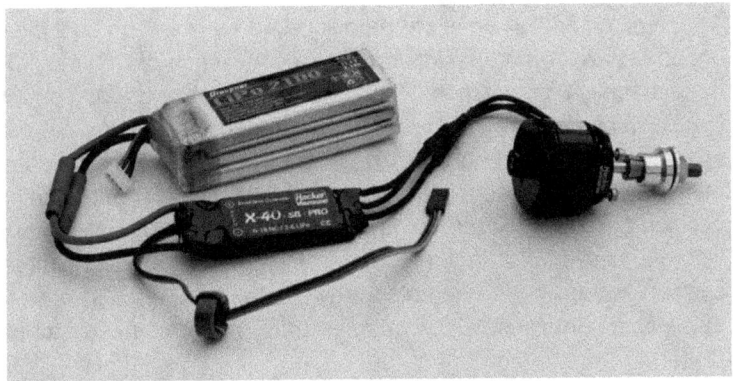

*Figure 3: Standard drive combination*

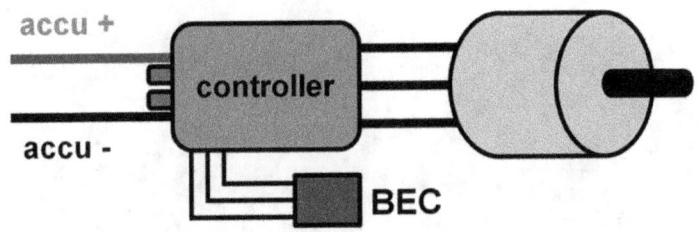

*Figure 4: Schematic illustration of the drive in Figure 3*

## 1.4 Brushless DC motor or just brushless motor?

The correct naming is almost a question of philosophy. The brushless motor is mentioned in the literature, sometimes abbreviated as BL motor and sometimes as BLDC motor. Depending on whether the motor is only meant for use by itself or in combination with the control, both terms are correct. The fact is, the motor itself has nothing to do with DC. DC means Direct Current. The simple designation brushless (BL) motor is then sufficient. If the whole motor and controller combination is meant, the name brushless direct current (BLDC) motor is quite reasonable. The combination actually appears like a normal DC drive. At the input of the regulator, the DC voltage of the battery is applied and at the shaft is mechanical power, just as one knows from brushed drives.

The term 'controller' needs to be discussed as well. The controller undertakes a lot more tasks than just a conversion of DC into AC, the electronic commutation. Rather, it is responsible for the correct operation of the drive in all respects. It is responsible for the correct speed of the motor and slows it down, if necessary. It also does much more, as described in Chapter 4: it must find the rotor position for control, has to recognize a potentially low battery voltage, or even reduce the speed in the event of an overly large current. In addition, it often also provides the receiver with power. The name 'Brushless DC Controller' has prevailed for all these tasks.

## 2. Working principle of the brushless motor

### 2.1 Active principle

Figure 5 shows a schematic diagram of a brushless motor. It is a so-called inrunner. Unlike the outrunner to be looked at later, the rotating part is inside and the fixed part is outside. The rotating part is called the rotor, here drawn with the permanent magnet in the middle, and the fixed part is called the stator. In practice, a motor would never be built in this way. But in order to understand the basic functions, this description is just right.

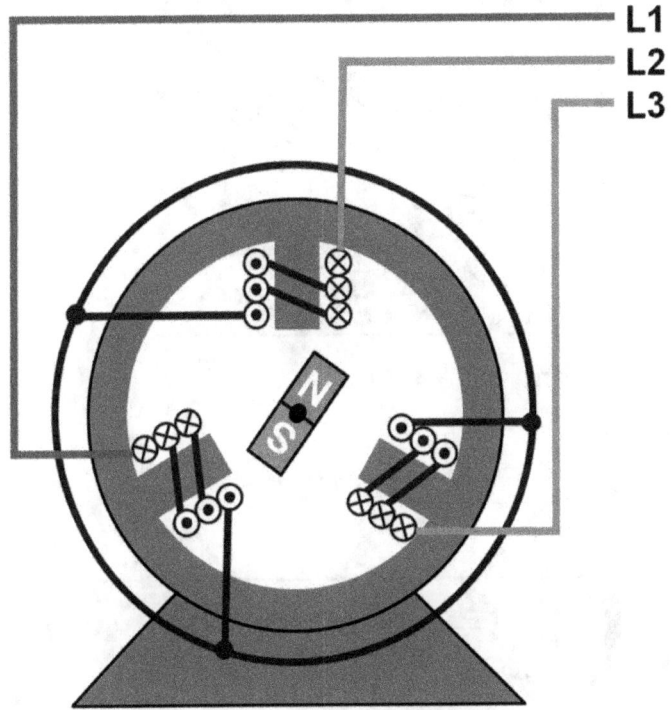

*Figure 5: Schematic diagram of a brushless motor*

Many properties can already be seen from this figure. If you were to ask what the biggest difference between a brushless and a brushed motor is, probably many would give the answer: "A brushless motor has three wires, while a brushed motor has only two". This is indeed true and can be seen here.

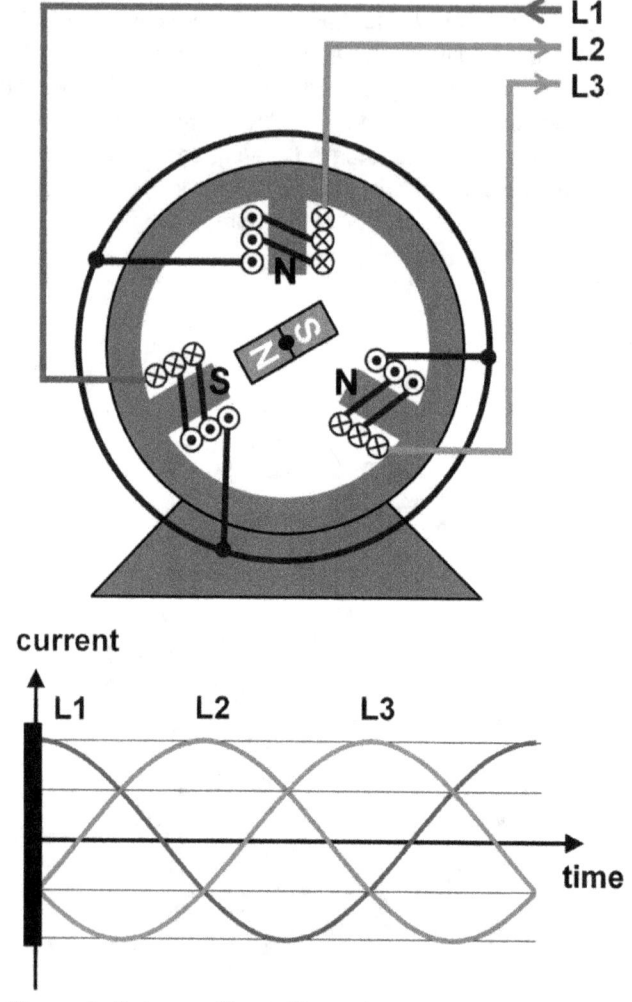

*Figure 6: Rotor position with maximum current through L1*

Each of the cables leads to one of the three coils of the stator. These are each arranged 120° relative to each other. The coil ends are all connected. This arrangement is called star connection; the interconnected coil ends are the star point. The delta connection is also possible, but this is described later.

Figure 6 shows the position of the rotor and, below that, the time course of the three currents that flow through L1, L2 and L3. It can be seen that they run sinusoidally. Exactly one full period, i.e. a course over 360°, is illustrated here. The three currents are displaced by 120°, in the order L1, L2 and L3. This can be nicely seen from the particular maximum values. If one imagines the time course now further, the maximum of L3 is followed by the maximum of L1 again. So one could string together the illustrated section of the curve ad infinitum. The bar represents the values of the currents at the present time. The current through L1 is located here just on the positive maximum, while the currents L2 and L3 are just on the negative half-maximum, and therefore flow in the other direction.

This fact can also be seen in the drawing of the motor. The current through the conductor L1 is pointing inward and the currents through the conductors L2 and L3 are pointing outward. The two arrows of L2 and L3 are illustrated more thinly because they are only half as large as the one of L1. With an inward-pointing current a magnetic south pole results at the coil on the stator. Accordingly, magnetic north poles are found in the coils, which are fed by the conductors L2 and L3. Because the south poles of the stator attract the north poles of the rotor and vice versa, the north pole of the rotor-side permanent magnet rotates exactly to the single south pole on the stator. The south pole of the rotor lies in between the two stator north poles.

Figure 7 shows the time bar a short time later. The current through L2 is now zero, and the two currents through L1 and L3 are almost at the maximum and have the opposite sign. Correspondingly a magnetic south pole develops in the coil which is fed by conductor L1. In the coil which is fed by conductor L3, a north pole develops. In the coil which is fed by conductor L2, neither north nor south poles are found. The permanent magnet rotor rotates so that its north and south poles are as close as possible to the south or north poles of the stator.

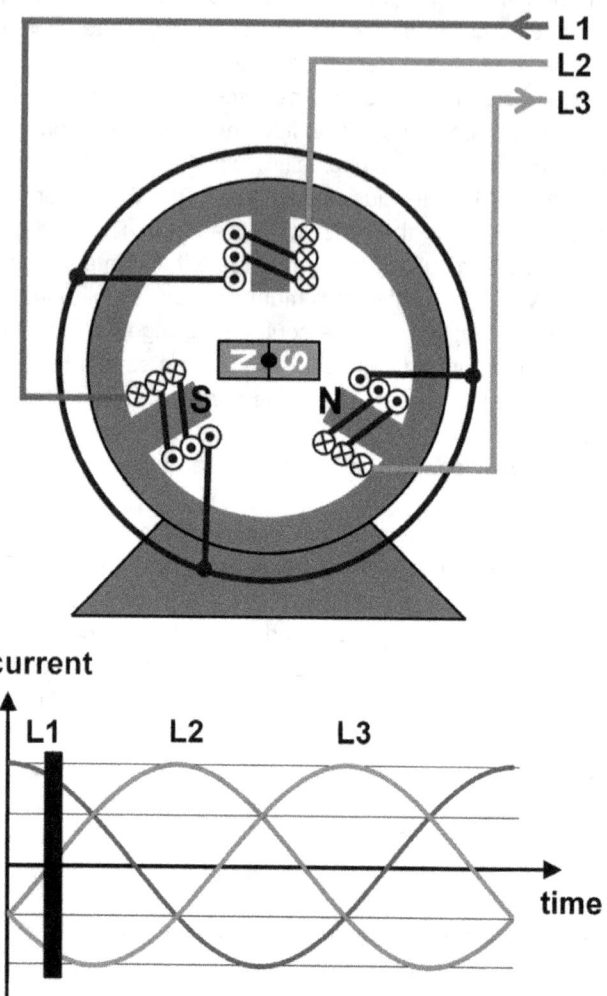

*Figure 7: Rotor position a short time later*

Figure 8 shows the time bar again a little later. It develops a magnetic north pole in the coil which is fed by conductor L3, analogous to the observations of the previous images. The other two coils are related to magnetic south poles. Accordingly, the permanent magnet rotor rotates a little further clockwise.

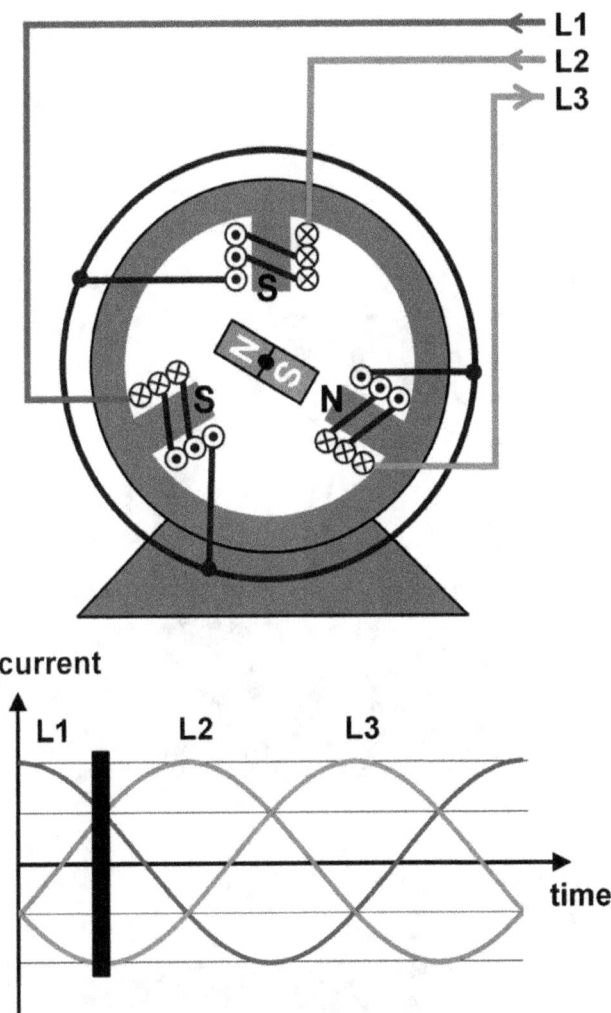

*Figure 8: Rotor position again a little later*

In Figure 9 the bar is such that the maximum current is flowing through L2. Through L1 and L3 are flowing the negative half maximum currents. Accordingly, a south pole and another two north poles develop. Compared to Figure 6, the time bar has now gone through a third of the period, therefore 120°. The rotor behaves in the same way. It has turned between Figures 6, 7, 8 and 9, exactly a third of a full rotation, i.e. the same 120°. If we consider the further motion of the bar, until it has completed a full period, then it means that the rotor has also rotated exactly once around its own axis.

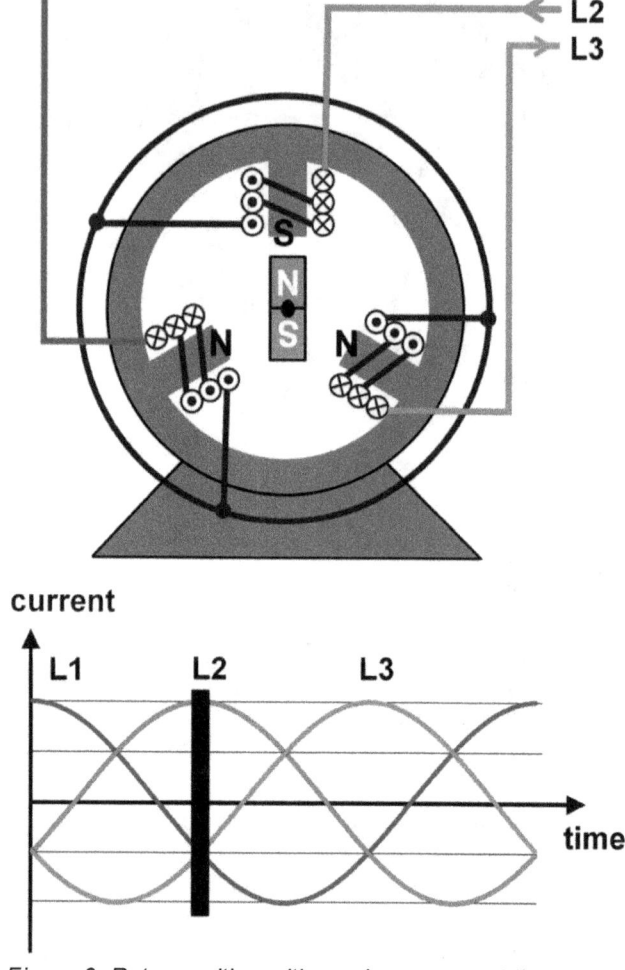

*Figure 9: Rotor position with maximum current through L2*

The rotor position and the position of the bar on the timeline of the current curve are coupled together forever. One can also say that the rotor is turning synchronously with the current curve. The name 'brushless motor' is now established as a term. But the invention is not new. The brushless motor is also found in the literature under the name 'synchronous motor'. It bears this name because of the correlation described above.

## 2.2 Action of force on a current-carrying conductor in a magnetic field

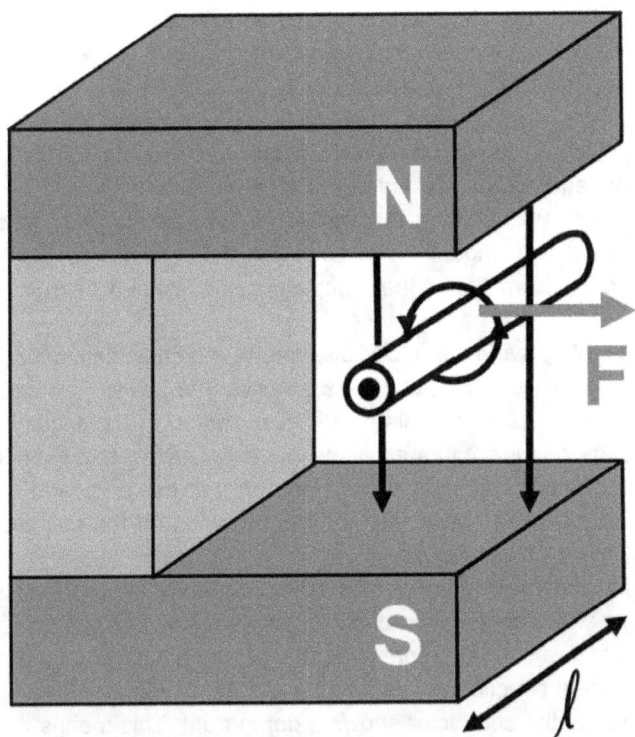

*Figure 10: Action of force on a current-carrying conductor*

Figure 10 shows an arrangement with a current-carrying conductor, which is located in the magnetic field of a permanent magnet. Because in Chapter 3 there will be a quantitative calculation, at this point a brief description of the force acting on a current-carrying conductor in a magnetic field is presented. The corresponding theories can be found in the literature under the terms 'Lorentz force' or 'Lenz's rule'.

The U-shaped permanent magnet has a north pole on the top and a south pole on the bottom. As with all magnets, so-called magnetic field lines exist. If one were to place this magnet on its side under a piece of paper and sprinkle iron filings onto it, they would collect in lines on the air link between the north and south poles. Two of these are shown in the figure.

Field lines are directed and self-contained, as noted in the literature. The direction is always from the north to the south pole; in the figure this is illustrated with the arrowheads. In fact, iron filings also collect along lines inside the U-magnet. The two lines drawn in the image therefore actually run on the inside of the permanent magnet, actually resulting in closed lines. However, they are not shown for the sake of clarity. For a better understanding, Figure 13 is however already referred to. At the inrunner shown there, the complete self-contained field lines are drawn.

A current-carrying conductor also shows magnetic field lines. They form closed circles around it, as shown in the image. The direction is now important, but this must be explained in more detail. The direction of current can either be illustrated by an arrow along the conductor or by an icon in cross-section. For this purpose Figure 5 should again be considered. There, the wires of the coil are also shown in cross-section.

A dot drawn in the cross-section symbolically represents an arrowhead. The direction of current is therefore 'out of the picture'. A cross symbolically represents the tail of the arrow. In this case, the current direction points 'into the picture'. In Figure 10 the cross-section of the conductor shows a dot in front. This means that the direction of current is from back to front.

A particular current direction is always coupled with the direction of the magnetic field lines. If we take the right hand and point with the thumb in the direction of the current flow, the four remaining fingers point in the direction of the field lines. In Figure 10, they thus point counterclockwise around the conductor. This correlation is also known under the term 'right hand rule'.

This fact will only be interesting with the formula which describes the force effect quantitatively. A force always acts in a certain direction. It is therefore also illustrated by an arrow. Also, the current and the magnetic field (the induction or flux density is here indicated as a calculation variable) are illustrated by arrows. Such variables are also called vectors. To calculate the force from the induction and the current, there is now a mnemonic:

*If the induction and the current are perpendicular to each other, the force is also perpendicular to both and has the value:*

$$F = B \cdot I \cdot L \quad \text{(Force = Induction x Current x Length)}.$$

L is the length over which the magnetic field acts on the conductor. As illustrated in the picture, the force points to the right and stands as stated in the mnemonic, vertically on both the current and on the induction. Even if the force were to point in the opposite direction, it would be perpendicular to the two other variables. So there would be two possibilities for its direction.

To find out in which direction it actually points, the following consideration can help. Left from the conductor, the two magnetic fields of permanent magnet and conductor are superimposed, as they point to the same direction. This means that the whole field will be bigger there. One can also imagine it that the field lines are denser there. The field lines would now like to hold the maximum distance from each other; they would like to use the space.

Figuratively speaking, they are now trying to push the conductor away to the right, to obtain more space. This fits well to the conditions on the right side of the conductor. There, the arrow of the magnetic field line caused by the conductor points upwards, while the arrow of the magnetic field line caused by the permanent magnet points downwards. The field is therefore small there, or otherwise expressed, the field lines are further apart there. With this in mind, the direction of the force is clearly established. It points to the right, as drawn.

## 2.3 Force action in practice on the example of an inrunner

The force action treated in the last chapter is the principle mode of action of most electric motors. To understand the functioning of the motors, we need to identify the permanent magnets which cause a magnetic field on the one hand and the current-carrying conductors on the other.

*Figure 11: Inrunner, stator with iron-free winding*

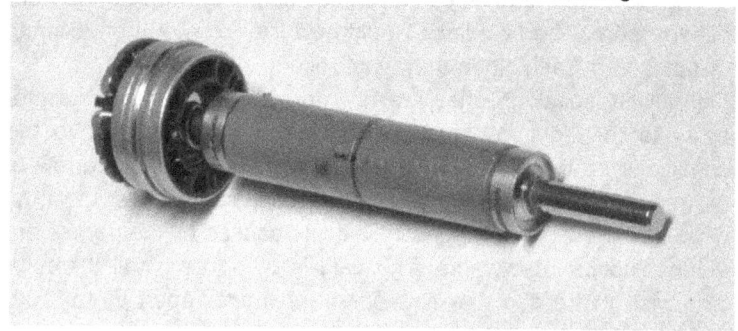

*Figure 12: Inrunner, rotor with permanent magnets*

Figure 11 shows a stator and Figure 12 a corresponding rotor. We are dealing here with an inrunner as in the previous chapter, where the active principle was described. The permanent magnet on the rotor is annular. If you approach it with a south or north pole of another permanent magnet, then you realize that when turned one side is repelled and the opposite is attracted by half a turn. The same was already shown in Figure 5, so there was also just one north and one south pole with this rotor. It is said in the jargon that this is a two-pole rotor or a rotor with one pair of poles.

Taking a closer look at the stator and comparing it with Figure 5, one can immediately see that the slots which contain the windings are missing. It seems as if the winding was laid as a carpet tile on the inside of the stator. In fact, this is a very frequent version of the inrunner, the so-called 'iron-less winding'. If the previously derived theory is to be applied here, those must be the current-carrying conductors, to which there is a force effect according to the laws of magnetism.

Figure 13 shows a schematic representation of the motor in Figures 11 and 12. The rotor also has one permanent magnet, as mentioned above. As in the last chapter, the field lines are self-contained as required. According to the rule that their direction points from the north to the south pole, they run to the top first. Then they follow the radius of the iron stator ring and come out of this to return to the south pole.

One could also imagine other paths of the field lines. These could, for example, even run within the rotor chamber from the north to the south pole. This idea, however, contradicts another property. The field lines, like many other technical values, also follow the path of least resistance. Air, of which the rotor chamber consists, presents a much larger so-called magnetic resistance than iron, which is a so-called ferromagnetic material.

On their way from the north to the south pole, the field lines therefore seek those regions in which they find as much iron and as little air as possible. In this way the field lines are formed as shown in the figure.

We turn now to the winding, which was previously a little disrespectfully called 'carpet tile'. If the active principle treated in Section 2.1 is also valid here, there must still be three windings present. That is true. The actual winding scheme will be ignored at this point.

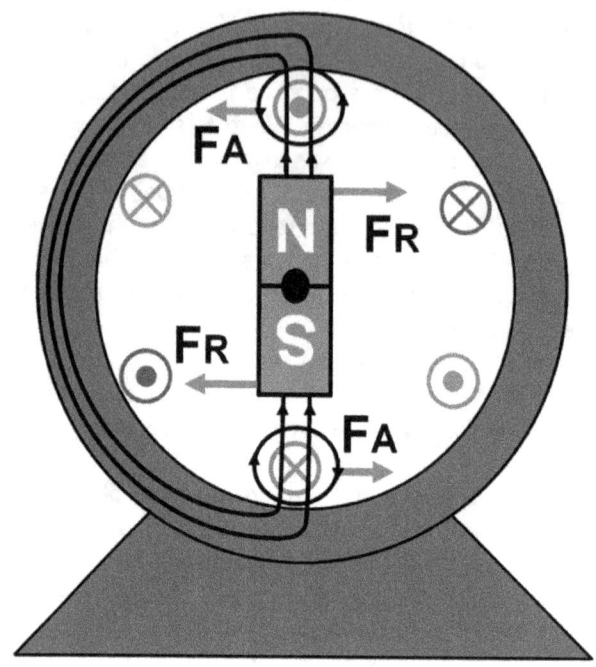

*Figure 13: Principle scheme of the inrunner of Figures 11 and 12*

But the coils shown in Figures 5 to 9 can also be found here. The coil L2, for example, leads along the length of the stator. Viewed in cross-section, the current comes out of the picture on the top and goes back inside on the bottom. Because of the rotor there is no room to lead the winding on the shortest path from top to bottom. Therefore it is guided along the inside of the stator. The result is that the windings are bent at the end of the stator. They are also called 'winding heads'. Of course there is no single winding, but several of them. This is so because the formula for expanding the force with the winding number N is as follows:

$F = B \cdot N \cdot I \cdot L$ *(Force = Induction x Turns x Current x Length)*

In Chapter 3, the turns N are treated a little deeper. They are a key item in many brushless motors and controllers. The field lines around L2 proceed according to the 'right hand rule'. On top they run counter-clockwise and at the bottom they run clockwise.

If the same considerations as in the last chapter are made, then at the upper conductor an enlargement of the magnetic field is developed on the right side, causing a force to the left onto the conductor. At the bottom conductor on the left are the same relations, which causes a force onto the conductor to the right.

The stator and its conductors can't move, however, since they make up just by definition the fixed part of the motor. In physics, action = reaction is taught. This is also relevant here. Every action force $F_A$ causes a reaction force $F_R$. So, if the stator can't move, the rotor must do so and, as shown in the figure, with a force onto the other side. This results in a clockwise rotation of the rotor. It is exactly the same as described in Section 2.1 in Figures 6 to 9.

As anyone who has ever held a starting (small! <10W) inrunner in their hand probably knows, for any action force there is also a reaction force. As will be discussed in Chapter 3, forces achieve torques in motors. At the start, when the rotor has a large torque, the stator moves with the same torque in the opposite direction. If you're lucky and the power is small enough, you can keep the motor in your hand. But the torque is felt. The explanation for this phenomenon is just the same as described above.

The true windings of the motor from Figures 11 and 12 are of course more complicated. But the curve of the field lines is also in reality as shown in Figure 13.

**Iron-free windings**

Iron-free (in other words coreless or slotless) windings have their advantages and disadvantages. With coils with a core, the windings lie in slots, and the production technology can keep the air gap between the rotor and stator iron very small. The smaller the air gap, the greater the induction or flux density B caused by the permanent magnets. In a coreless winding the air gap needs to be selected somewhat larger, since the coil comes to lie in it. Moreover, this manufacturing technology can't be executed in tight tolerances, which increases the air gap even further. For this reason, the induction B is typically not as high as for a motor with slots. But it scores points in the so-called cogging torque. The shaft of the currentless motor with a coreless winding can be turned pretty equally, because the whole circumference is symmetrical. This increases the efficiency. A disadvantage is the relatively long winding head, which is bent at least on one side around the circumference of the stator, to ensure that the rotor can be installed at all. Only on the length L does the winding contribute to the force effect, and not on the winding head itself. This, however, enlarges the winding resistance, which reduces the efficiency. This is why motors with iron-free coils are usually made thin and long. The one illustrated in Figures 11 and 12 is therefore a typical example.

### Properties of inrunners, gearbox

The brushless inrunners have similar characteristics as DC (brushed) motors. These are – with the exception of special designs – almost always inrunners. Both motor types are subject to the same principle of force acting on a current-carrying conductor in a magnetic field. Both show rather a high speed at a relatively small torque. But in most applications, such as a propeller-driven model aircraft or the wheel drive of cars, a lower speed with a larger torque is required. Thus, these motors are generally used together with a gearbox. The speed is thus smaller and the torque greater. The available mechanical power at the drive shaft is calculated without gearbox by

$$P_{motor} = 2 \cdot \pi / 60 \cdot n_{motor} \cdot M_{motor}$$

where $n_{motor}$ is the rotational speed in revolutions per minute and $M_{motor}$ is the torque at the motor shaft in Newton meters. With a gear ratio i the formula is as follows:

$$P_{gearbox} = 2 \cdot \pi / 60 \cdot n_{motor} / i \cdot M_{motor} \cdot i$$

Here $n_{motor} / i$ is the speed and $M_{motor} \cdot i$ is the torque at the gear shaft. In the formula, the gear ratio i is canceled down. An i of 2, for example, is achieved with a cogwheel at the drive shaft (pinion gear) with 15 teeth and a gear on the transmission shaft (spur gear) with 30 teeth. Gearboxes have efficiencies that are approximately 0.9. Considering this, the gearbox efficiency is multiplied by the motor efficiency (see Chapter 3) and deteriorates to the efficiency of the overall drive by about 10%.

## 2.4 More stator slots and magnetic poles on the examples of outrunners 6S8P and 12S14P

It would often be desirable for model constructions to have a drive with lower speed and higher torque. A reasonable drive should therefore unite all these advantages. It should on the one hand be 'brushless' because of the high efficiency and few wearing parts. On the other hand it would be nice if a mechanical gearbox could be omitted, and the motor would exhibit at the shaft a large torque at low speed.

Figures 14 and 15 show a brushless outrunner motor. It is easy to see that here the stator with the windings in the slots, 12 in number, is inside. The rotor is pushed with the shaft into the bearings of the stator. It looks like a bell and on the inside carries small magnets, 14 in total. They are alternately lined up as north and south poles. This motor is a so-called 12-slot-14-pole, a 12S14P. It is a very commonly produced type.

But why should such a motor have a higher torque than an inrunner, one might wonder. For this purpose a rough calculation is made. The force which acts on the rotor is enlarged for a motor with increasing diameter. Since then also the slots are larger, one can also bring up more copper and thus more windings. In addition, there is torque = radius x rotor force. Both rotor force and radius are larger. Thus, it is easy to see that the torque is increased strongly if the permanent magnet is far away from the center. The outrunner is therefore the ideal form for a motor with high torque.

Also its production is relatively simple. One sticking point in the manufacture of brushless motors is the mounting of the permanent magnets. One usually overcomes this with epoxy glue. With inrunners, the centrifugal forces threaten to tear apart the rotor magnets. With outrunners, they are just pressed more against the rotor bell. An ingenious design!

*Figure 14: Stator of an outrunner*

*Figure 15: Rotor of an outrunner*

In the first years of the 21st Century, there were relatively few brushless motors on the market. There were many model constructors who built their own motors. In CD and DVD drives, brushless outrunners have been standard for many years. They are among the first products for which these drives were produced in large numbers. So people took out the old drives and used the stator slots to build their motors. At this point the author wants to limit the description to two typical outrunners, the 6S8P and 12S14P, shown in Figures 14 and 15.

**The 6S8P outrunner**

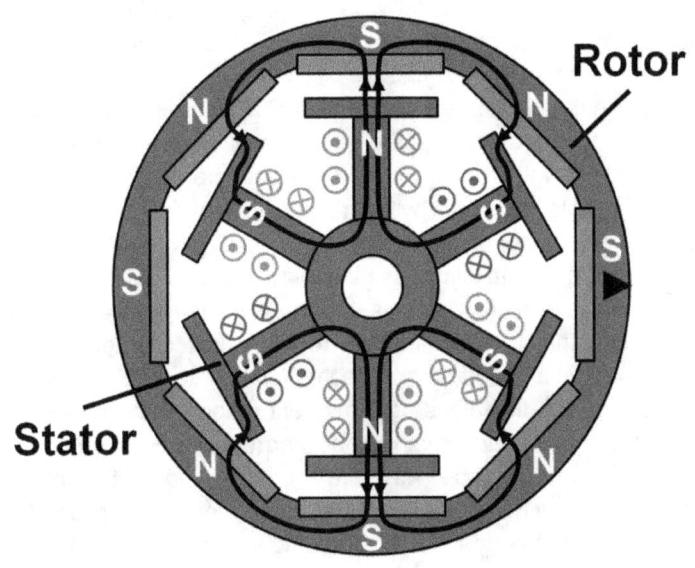

Figure 16: 6S8P outrunner

This motor is also called mini-LRK. LRK are the first letters of the surnames of Christian Lucas, Ludwig Retzbach and Emil Kühfuss. Turning again to the question of how many stator slots are useful for a brushless motor with three coils, you can find out easily that it must always be a multiple of three. According to Section 2.1, there are three wires which lead to the motor. To keep things symmetrical, this must be the only conclusion. In model construction today, almost all types of integer multiples of three are built, from six stator slots up to about 27 stator slots, in many different types and power classes.

In line with the above description, the 6S8P has six stator slots and eight permanent magnet poles on the rotor. Figure 16 shows it in cross-section with the associated current/time graph. Of course, also this motor operates with the principle of force acting on a current-carrying conductor in a magnetic field.

It would be quite difficult to use this principle to understand such a motor, because more windings and more poles are available than in the example of the inrunner from the last chapter. For this reason, here a different approach is taken to explain what is happening. The field lines of permanent magnets and those of all current carrying conductors overlap and there are new field lines resulting from this. This was already discussed in Section 2.2. There the resulting field lines in Figure 10 were denser left of the conductor but less dense right of the conductor, causing the force effect on it. The action of force can be understood with the reversed idea that the resulting field lines on the one hand want to minimize their distance which they cover in the air, in order to minimize the magnetic resistance, and on the other hand they want to establish the maximum possible distance to the next field line. To achieve all this, they push away the conductors with force.

In the situation in Figure 16, through conductor L2 the maximum current is flowing whereas through L1 and L3 the negative half-maximum current is flowing. This results in electrical north poles in the upper and lower stator yoke (the name given to the area where the iron and not the slot is). The four yokes upper left, lower left, upper right and lower right obtain electric south poles from the half intensity.

These poles now attract the north respectively south poles of the rotor. This is also illustrated with the field lines. By the two strong north poles of the stator lead two field lines. They split up in the rotor iron, and one returns by the left and one by the right neighbor yoke, to the respectively half as strong south pole.

There a balance is found; the rotor turns neither left nor right. But if you want to turn it away on one side, then the stator exerts a force on it and wants to retract it again.

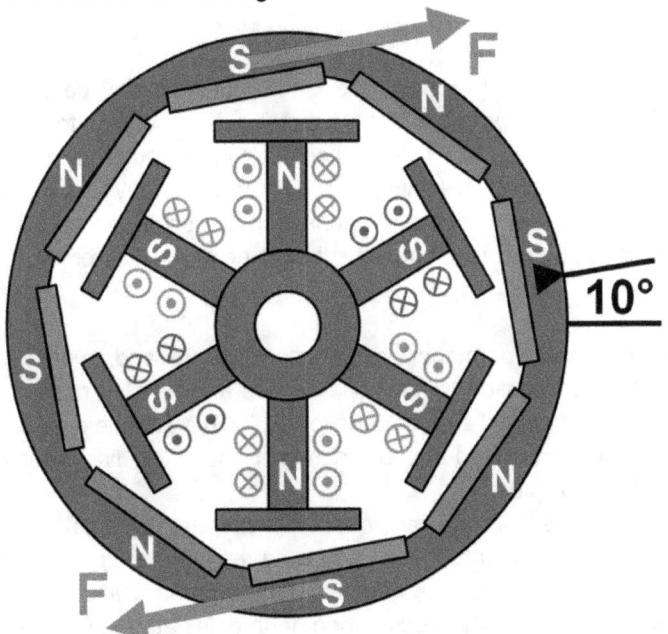

*Figure 17: 'Lagging behind' rotor*

Figure 17 shows a state in which the current/time diagram of Figure 16 is already valid, but where the rotor still lags behind at 10°. The magnetic field lines have not yet minimized the air gap, and therefore the stator attracts the rotor in a clockwise direction. The two drawn-in forces in south poles at the top and bottom are the strongest, but also the other rotor magnets are attracted by forces in the same direction. In order that the motor can deliver any torque at all, the rotor needs to always be slightly lagging behind.

Once it's no longer lagging behind, as in Figure 16, the forces and the torque are zero. There is the following mnemonic:

*The rotor of a brushless motor rotates synchronously with the curve of current. If it produces a torque, it lags slightly behind to the curve of current. The greater the torque, the stronger it lags behind the AC current curve.*

If the torque is too large, the lagging behind may be too large. Then it is also possible that the motor gets uncoupled from the AC current curve. One says then that the motor falls out of synchronization. This should not happen under any circumstances. Therefore, the brushless DC controller needs information about the rotor position. This will be described in detail in Chapter 4.

With so many permanent magnets on the rotor and so many stator slots, the question is now whether the rotor is also making a full turn during an AC current curve of 360°. This was the case with the inrunner considered in the last chapter and in the discussion of the active principle.

Figure 18 shows the motor with the present currents 120° later. The maximum current now is flowing through the coil L3; through the coils L1 and L2 the negative half-maximum current is flowing. Accordingly, this develops strong north poles at the two yokes with the coil L3. At all other yokes less strong south poles are resulting. In all the examples which have been treated so far, the rotor was rotated at the same angle as the AC current curve. If one compares Figures 16 and 18, and looks at the triangular mark on the right side of the rotor, it is illustrated that the rotor now rotated only 30°. The AC current curve on the other hand has rotated by 120°. The field curves are again exactly the same (just turned by 120°), but in contrast to the above examples, not the same south poles of the rotor are attracted by the new north poles of the stator, but those which were already before next to the stator with the coil L3.

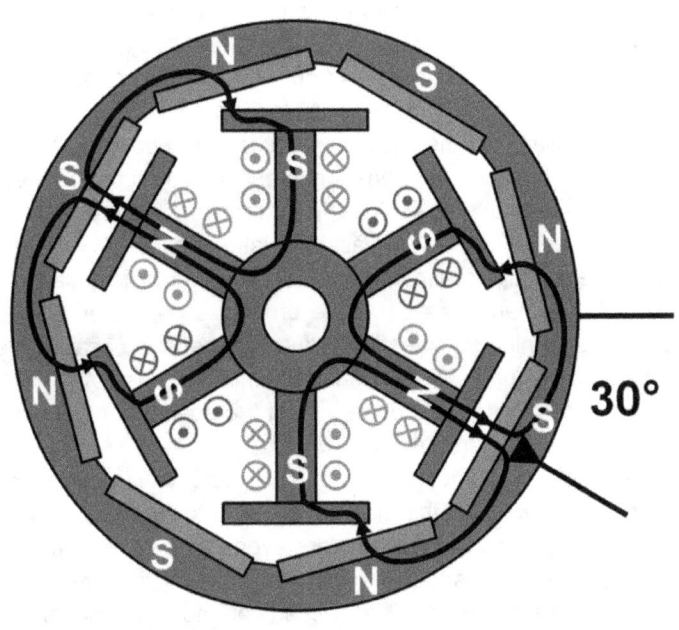

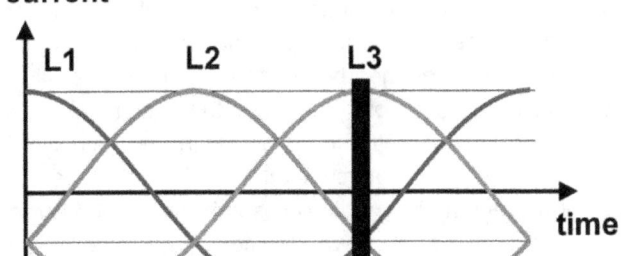

*Figure 18: 6S8P outrunner, AC current curve 120° later*

This reduction ratio can be detected with a formula:

$$i = \frac{\text{current frequency}}{\text{rotational frequency}} = \frac{\text{permanent magnet poles}}{2}$$

Applied to the 6S8P a reduction ratio of 4 results. So that the motor can rotate once around its axis, the AC current has to go through a total of four periods, equal to 4 x 360°. This formula applies also to the schematic diagram of the brushless motors given in Section 2.1, and similarly for the inrunner discussed in Section 2.3. These motors have only one permanent magnet, with a north and a south pole, so equal to two poles. Their gear ratio is therefore 1, and the current frequency is the same as the rotation frequency.

The above formula is also on the second look almost sensational in terms of the characteristics of the brushless outrunner. Because the rotor is running outside, the torque is larger and because many permanent magnets are seated on the rotor, due to the larger diameter, the rotational frequency (and the speed) is even smaller than the AC current frequency.

This is an electronic gearbox. With the special drives designed for the application it would therefore be possible to avoid a mechanical gearbox altogether. Brushed motors even have a lower efficiency because of the 'brushed' attribute. In addition, if a mechanical gearbox is added, which is the case for most applications, the efficiency decreases again by 10%. In addition, the mechanical gearbox also brings additional weight on the scale. All this together makes up the success of the brushless outrunner.

However, this electronic gearbox can't be increased easily to any level of reduction. On the one hand, due to the higher frequency of the AC, losses occur in both the motor and in the controller. On the other hand, in multi-pole rotors leak fluxes also occur, because the density of the alternating north and south poles will be too large and this also results in unwanted field lines directly between neighboring north and south poles.

But in practice, the mechanical gears in the model construction are usually between 1:2 and 1:20. This can more or less be realized with the electronic gearboxes of brushless motors.

### The 12S14P outrunner

It is beyond the scope of this book to discuss all the variations of outrunners. So at this point only a very commonly observed representative, the 12S14P of Figures 14 and 15, will be discussed.

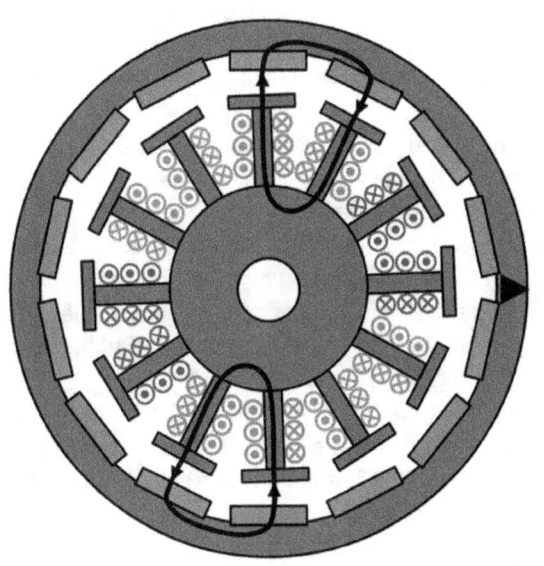

*Figure 19: 12S14P outrunner, maximum current through L2*

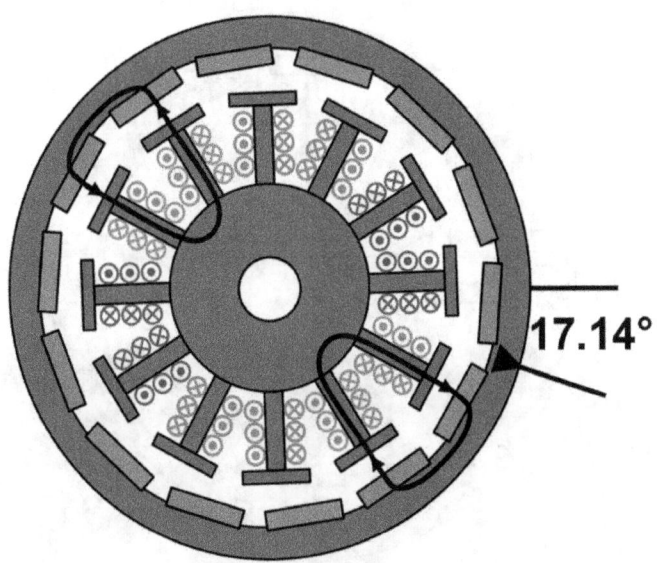

*Figure 20: 12S14P outrunner, maximum current through L3*

The standard coil is the so-called distributed winding. It is presented in Figure 19. A conductor is wrapped with the above winding direction around two neighboring stator yokes before it is wrapped around the two opposite yokes. The four sub-windings constitute the entire coil. Thus results in a north and a south pole on each side. The field lines close through the south and north poles of the permanent magnets on the rotor.

Drawn in a simplified manner, Figure 19 only shows the field distribution of the wire L2, which in this example leads the maximum current. Again, it is once more the stable case in which the rotor can turn neither to the left nor to the right. Recall that the rotor is slightly lagging behind under load.

In Figure 20, the same motor can be seen 120° later. Now the wire L3 is leading the maximum current. The gear ratio is based on the above formula 14 / 2 = 7. If the AC current turns by 120°, then the rotor rotates just 120° / 7 = 17.14°.

Again, only the course of the field lines, caused by the current through L3, was marked. It is noticeable, however, that the field lines close here only on the yokes of their own coil.

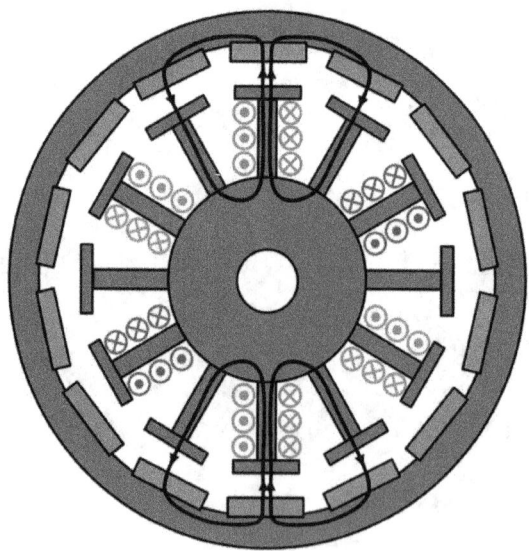

*Figure 21: 12S14P outrunner with LRK windings*

But an LRK winding also exists for the 12S14P. It can be found in Figure 21. The difference from the distributed coil is that only every second stator yoke is wrapped. At first glance one might think that this is a waste of iron, because the unwrapped yokes would contribute nothing at all. But of course this is not so, because the unwrapped yokes are important for guiding the magnetic field.

The field lines close again as in the mini-LRK coil. They are divided in the rotor and run back through the left or right yoke respectively. The unwrapped yokes also lead back the field lines from the other neighboring coil (not shown here).

**Cogging torque**

In the last chapter an inrunner with an iron-free or coreless coil was considered. It was mentioned there that they have no cogging torque due to the symmetry. Of course there also exist inrunner motors with coils in slots. All the outrunners considered above have these as well. In this case, the permanent magnets and the stator yokes will attract each other in the current-less case as well. If the permanent magnet is just above the yoke, as can be seen for example in Figure 16, it doesn't like to move away from it. He who wants to turn the rotor of such a motor by hand, notes that he can't rotate equally, because the rotor is attracted during the turn again and again. During normal operation when the drive delivers power, the cogging torque is superimposed on the motor torque, which is constant over the entire revolution.

It turns slightly faster when the permanent magnet is close to the yoke and slightly slower when it is located just after the yoke. However, this torque ripple also operates in reverse in accordance with Section 2.2, in that the current in the winding and the magnetic flux in the iron varies slightly. This leads to slightly larger losses and reduces the efficiency.

## 2.5 Star connection, delta connection

In the motor drawings on the previous pages, there are three windings, which are connected to the three supply cables.

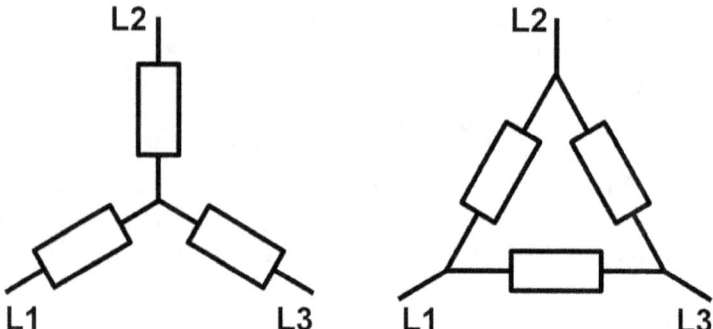

*Figure 22: Star connection (left), delta connection (right)*

But a coil has two ends, and three coils give a total of six ends. To combine these six ends meaningfully with the three supply cables, in Figure 22 the two possibilities of the star connection or the delta connection are shown. With the star connection, one end of the coils is always connected to one of the supply cables. The other three ends are all connected together. This connection point is called the star point. It is usually not accessible from the outside. In the second winding type, the delta connection, the ends of two coils are connected together with a supply cable.

When discussing the differences between the two types of connection, must first be answered the question whether it is guaranteed in every case that the three coils lead by 120° phase-shifted currents. Above, this was indeed the requirement for a proper function. A discussion of Figure 22 clarifies this question. The three supply cables of the signals phase-shifted by 120° can also be illustrated visually, so that they form an equilateral triangle. Through symmetry, the result for both the star and the delta connection is that the three coils in the figure are also oriented by 120° to each other. But that means in the reverse reflection that the voltages and currents of these coils are also oriented by 120° to each other. The proper function of the brushless motor is thus ensured in both cases.

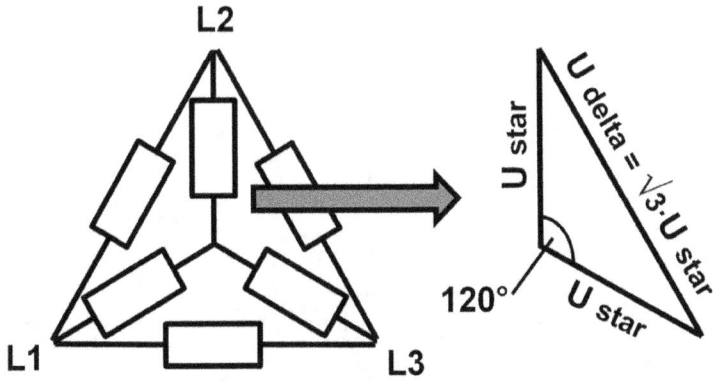

*Figure 23: Voltages with star and delta connections*

The two types of connection, however, are not equivalent. Figure 23 (left) shows them drawn over each other. This results in three isosceles triangles with a 120° angle between the arms. The two arms in this case represent the voltage over a coil in a star connection. The third side represents the voltage of a coil in a delta connection. With a bit of trigonometry it can be found out that the third side is greater than the two arms by √3. This yields the following formula:

$$U_{delta} = \sqrt{3} \cdot U_{star}$$

How this exactly affects the properties of the motor is covered in Chapter 3.


**Winding schema**
In a motor with lots of slots, the three coils are distributed to them. This can be seen in both examples – the 6S8P in Figure 16 and the 12S14P in Figure 19. Figure 24 shows how the windings of slots are interconnected.

39

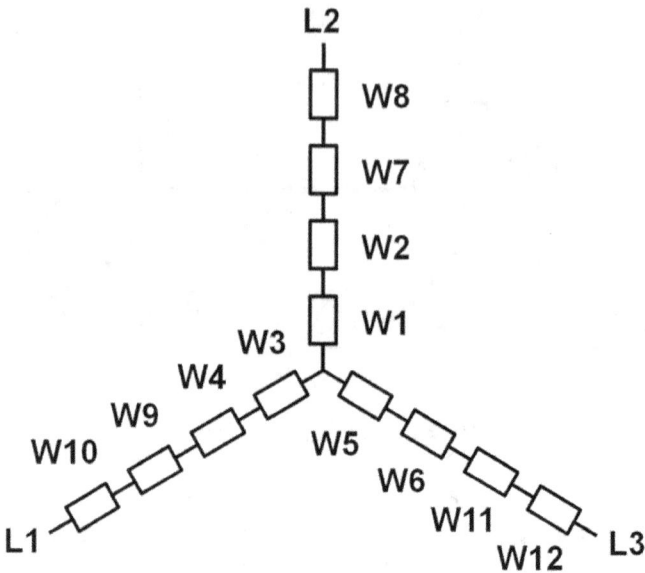

*Figure 24: Winding schema of 12-slot stator in star connection*

As an example a 12-slot motor in star connection is illustrated. The different windings are represented as resistors (more generally: impedances) and are here connected in series. The winding direction must be scrupulously respected, i.e. whether it is needed to be wrapped in a clockwise or counterclockwise direction. One can find this out for example through an illustration corresponding to Figure 19. The conductors with a point in the cross-section run out of the picture and those with a cross run into the picture. Alternatively, you can take a winding table to hand.

**Winding table**
There are many possibilities for windings. Often they are also equivalent to each other. Many of these and comparisons with one another can be found on the web. At this point, however, a short table of common types will be shown and explained. Often, the windings presented above as L1, L2 and L3 are written in the table with the letters A, B and C. This applies to a clockwise winding. For a counter-clockwise winding the small letters a, b and c will be

used. The two orientations can be reversed. Thus it is not important how you hold the stator in the orientation determination. It must then simply always be kept the same. It is not important whether the slots are counted clockwise or counterclockwise. This only has an influence on the direction of rotation. This can simply be changed by interchanging two of the three connecting cables.

6-slot stator (example 4-pole motor or 8-pole motor)

| stator yoke | 1 | 2 | 3 | 4 | 5 | 6 |
|---|---|---|---|---|---|---|
| winding | A | B | C | A | B | C |

This means that the coil A will start at yoke 1 and will continue thereafter with the same orientation in yoke 4. It is equal to coil B (yokes 2 and 5) and coil C (yokes 3 and 6).
This results in a wire end at each yoke; in total there are six. The following continues to apply:
Star connection: connection of wires 1,2 and 3. Junction: 4,5 and 6
Delta connection: connection of wires 1 and 6, connection of wires 2 and 4, connection of wires 3 and 5. Junction: the three connections

9-slot stator (example 12-pole motor)

| stator yoke | 1 | 2 | 3 | 4 | 5 | 6 | 7 | 8 | 9 |
|---|---|---|---|---|---|---|---|---|---|
| winding | A | B | C | A | B | C | A | B | C |

Also here there are six wires left at the end, at yokes 1, 2, 3, 7, 8 and 9. At yokes 4, 5 and 6 no wire is left
Star connection: connection of wires 1,2 and 3. Junction: 7,8 and 9
Delta connection: connection of wires 1 and 9, connection of wires 2 and 7, connection of wires 3 and 8. Junction: the three connections

12-slot stator (example 10-pole motor or 14-pole motor)

| stator yoke | 1 | 2 | 3 | 4 | 5 | 6 | 7 | 8 | 9 | 10 | 11 | 12 |
|---|---|---|---|---|---|---|---|---|---|---|---|---|
| winding | A | a | b | B | C | c | a | A | B | b | c | C |

The six wire ends can be found at yokes 1, 3, 5, 8, 10 and 12

Star connection: connection of wires 1, 3 and 5. Junction: 8, 10 and 12
Delta connection: connection of wires 1 and 12, connection of wires 3 and 8, connection of wires 5 and 10. Junction: the three connections

The winding scheme of the 12-slot LRK motor will be:
12-slot stator LRK (example 10-pole motor or 14-pole motor)

| stator yoke | 1 | 2 | 3 | 4 | 5 | 6 | 7 | 8 | 9 | 10 | 11 | 12 |
|---|---|---|---|---|---|---|---|---|---|---|---|---|
| winding | A | - | b | - | C | - | a | - | B | - | c | - |

The six wire ends can be found at yokes 1, 3, 5, 7, 9 and 11

Star connection: connection of 1, 3 and 5. Junction: 7, 9 and 11
Delta connection: connection of 1 and 11, connection of 3 and 7, connection of 5 and 9. Junction: the three connections

**Skin effect**

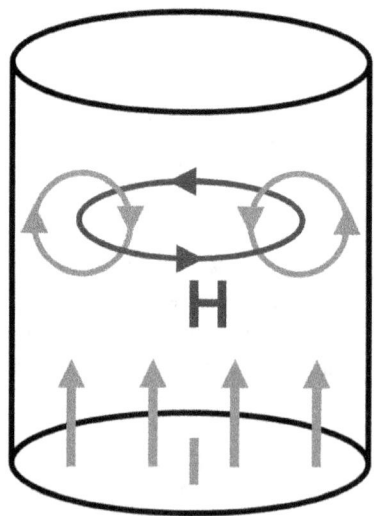

Figure 25: Skin effect

In many motor coils, each winding is not only one wire, but will be done with several in parallel. One may wonder whether it would be easier to take a single thicker wire which has the same cross-section as all parallel-connected wires together. A sufficiently large cross-section is important to reduce the resistance and thus the winding losses.

The skin effect makes sure that with AC current, several thinner wires in parallel have a smaller winding resistance than a single thick wire. Figure 25 shows what happens inside a conductor.

Current causes a magnetic field not only outside but also inside the conductor, as shown in Section 2.2. An AC current also causes an alternating magnetic field and this alternating field again causes (they also say in the technical language 'induces') a small AC current in the conductor.

Now, if the main current shown at the bottom of the picture and the induced currents are superimposed on each other, we see that the arrow directions are opposite to each other in the middle of the conductor. Further out, they point in the same direction. In the middle the current is so attenuated; outwards it is enlarged. It is also said there is a current displacement outwards.

This means that with AC only the outer layers of the conductor contribute to the current flow and the center contributes nothing or much less. Since not the entire cross-section of the conductor contributes to the current flow, the winding resistance relevant for the further calculations is also larger.

A measure for the current displacement is the so-called 'penetration depth' δ. It describes in which thickness, measured from the outer side of the conductor, most of the current flows. There are several parameters that influence δ. It is important to note that in addition to the material, also the frequency of the alternating current plays a role. For the copper conductor material the following values can be found:

f = 50 Hz: δ = 9.4 mm
f = 1 kHz: δ = 2.1 mm
f = 10 kHz: δ = 0.66 mm

One might think at first glance that this is not relevant for small electric motors, since the wires are thin and the frequencies not as high. For the context in Section 2.4, a rough calculation will be undertaken. As an example the 12S14P outrunner from Figure 19 is used. A typical representative makes at 10V approximately 10,000 rpm. Then the frequency of current is calculated as f = 10,000 / 60 x 7 = 1167 Hz. 10,000 must be divided by 60, because the frequency is defined 'per second' and not 'per minute'. The alternating current oscillates in a 14-pole outrunner 14 / 2 = 7 times faster than the rotor because of the electronic gearbox, so the result must be multiplied by 7.

The depth of penetration is therefore approximately 2 mm in comparison with the above values. In the case of a wire diameter of assumed 0.4 mm, one would notice this and it is therefore better to wrap in parallel several mutually insulated conductors with a smaller cross-section. If a single wire is replaced by several parallel thinner ones, the sum of their cross-sectional areas must be equal to the cross-sectional area of the single wire. Since the cross-sectional area is proportional to the square of the diameter, the table below applies. It is assumed that the single wire has a diameter of 1 mm. For larger or smaller diameters increase or decrease the diameters of more conductors proportionally.

| 1 wire | Replacement by 2 wires | Replacement by 3 wires | Replacement by 4 wires | Replacement by 5 wires |
|---|---|---|---|---|
| 1 mm | 0.71 mm each | 0.58 mm each | 0.5 mm each | 0.45 mm each |

Of course, the diameters of enameled copper wires delivered from stock are not available in increments of hundredths of a millimeter. Conventional sizes are in steps of five-hundredths, i.e. 0.1 mm, 0.15 mm, 0.2 mm, 0.25 mm, etc. To achieve a small winding resistance, it is wise to round up the calculated value in each case to the next thicker available diameter and to hope the winding is then not too thick for the slot.

## 2.6 Findings in brief

For a better understanding, the key findings from Chapter 2 will be summarized again.

*A brushless motor has three cables. These lead alternating currents, which are phase shifted by 120° to each other. The rotor rotates in synchronism with the frequency of these alternating currents.*

*The brushless motor is based on the principle of force on a current-carrying conductor. It follows the formula: $F = B \cdot I \cdot L$, where B is the induction caused by the permanent magnet, I is the current and L is the length over which the magnetic field acts on the conductor.*

*The speed of the rotor corresponds to the frequency of the alternating currents divided by the factor 'permanent magnet poles / 2'. In particular, outrunners with many poles have in this way integrated an electronic gearbox. This reduces the speed and increases the torque. For many applications this is ideal.*

*The success of brushless motors is based on very high efficiency and low wear. This has two main reasons:*

- *a.) The brushes, and thus the biggest wearing part of the DC motors, are eliminated. This also eliminates the efficiency-reducing friction between the brush and rotor.*
- *b.) The electronic gearbox of multi-pole rotors often replaces an efficiency-reducing mechanical gearbox, as speed and torque is adapted to the application at the motor shaft already.*

*There are two types of windings for a brushless motor: the star and the delta connection. They are not equivalent. In a motor with a delta connection, the voltage over the winding is by $\sqrt{3}$ larger than in a motor with a star connection.*

*To carry out the coils, one uses a winding schema or tables. It must be ensured that the specified winding direction is followed up. To avoid the skin effect, several thin wires are often wrapped in parallel instead of a single thick wire.*

# 3. Characteristics of the brushless motor

In the last chapter, the working principle of brushless motors was discussed with the focus on the basics, without calculation examples. Today there are many manufacturers of a variety of drives. Each customer must select the most appropriate one for his application. As assistance, in the data sheets a few characteristics of the motors are always listed. The most important ones are the revolutions per minute and volts (kV), a proposal for the number of (LiPo) cells, current consumption and efficiency. For a start, with those values the power at the shaft can be calculated. When the data sheets lead a bit further, also the winding resistance, the turns, the current consumption at maximum efficiency and the current consumption at idle running or warming are listed.
This Chapter discusses these characteristics based on the findings of Chapter 2. On the one hand, the relations between the different values will be shown, and on the other hand examples of drives from experience are calculated and verified by measurements.

## 3.1  kV, rpm/V

One of the most important characteristics is the one which is known by the names in the title. The two terms mean the same thing: how large the motor idle running speed is per volts. rpm/V means 'rotations per minute per volt'. The shorter term, kV, simply means a gain factor k for the voltage in volts. The unit is the same as with rpm/V. From the last chapter it is known that the supply currents and thus the supply voltages are AC. However, there is the question of which voltage could be meant now. The speed of a motor is mostly constant in operation, i.e. it has no alternating value. In Chapter 1, however, it was already mentioned that with brushless technology the combination of motor and controller is always a unit. With 'per volts' the DC voltage is meant, which is applied to the input of a lossless adopted controller. That is typically the accumulator voltage.

### 1st example:

The brushless controller of a Robbe Roxxy 2827-34 motor with kV = 760 rpm/V is applied to a 3S LiPo accumulator. '3S LiPo' means that three cells are connected in series, each with 3.7V. This results in a voltage of 11.1V at the input of the controller. The idle running speed (without load) is calculated as n = 3 x 3.7V x 760 rpm/V = 8436 rpm.

The motor will not quite reach this speed, because the friction losses in idle running also create a small load torque, which generates an idle current. However, for now this calculation is sufficient. The exact relations between the current flow and the speed are discussed in Sections 3.2 and 3.3.

### The induced counter voltage $U_{EMF}$

Strictly taken, 'per volt' is not for the voltage at the input, but for counter voltage $U_{EMF}$ induced from the rotating motor.

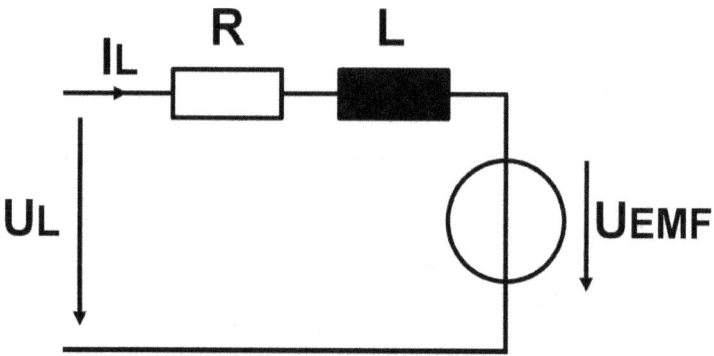

*Figure 26: Equivalent circuit of one of the three coils*

Figure 26 shows an equivalent circuit of one of the three coils, the so-called single-phase equivalent circuit. On the left side the conductor voltage $U_L$ can be found. The resistance R and inductance L represent the winding itself. Interesting is, however, what can be found on the right side, the induced counter voltage $U_{EMF}$. 'EMF' stands for Electro Magnetic Force. Obviously the spinning rotor produces a voltage itself, which appears in the equivalent circuit. In the ideal idle running, when no friction is present, conductor current $I_L$ would be 0. But this is only the case when the conductor voltage $U_L$ is of equal size as the induced counter voltage $U_{EMF}$. The calculation of the parameter kV reveals the secrets of this counter voltage.

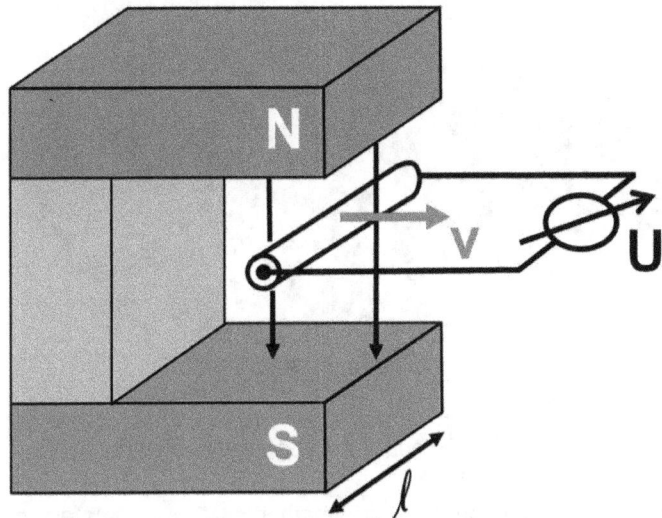

*Figure 27: Induced voltage*

Figure 27 describes the induced voltage of a conductor in a magnetic field. This arrangement was already shown earlier. That case concerned there the force on a current-carrying conductor. When an electrical conductor is moved through a magnetic field with the speed v, the result is an electrical voltage. It is then said 'a voltage will be induced'. For this effect, the same theories are again responsible as above. The voltage U is calculated according to the formula:

**U = B · N · v · L**   (*Voltage = Induction x Turns x Velocity x Length*)

This formula will be applied hereinafter to a real existing motor. The Hacker A20-22L has a kV of 924 rpm/V (read from the datasheet). Figure 28 shows the 12-slot stator of such a motor.

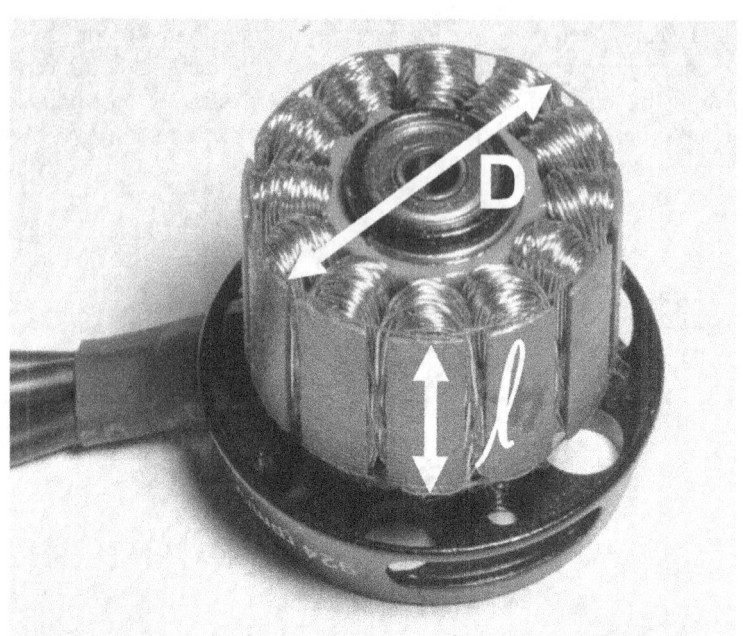

*Figure 28: Induced voltage in a real motor*

There is first the question of whether all the parameters of the above formula can be identified in this motor. U is the induced voltage, this is simple at first. B is, as already mentioned above, the induction which is generated by the permanent magnets. To perform calculations, this value must now also be known. Modern magnets are made with the chemical element neodymium (Nd). They are also called 'rare-earth magnets'. The value of the induction is up to 1.4T (Tesla). This is also assumed for the calculation. N is the number of turns. In this example, 22 turns per coil can be counted. The length L is drawn as the length of the stator, which is about 13 mm. Since the wire contributes on both sides to the coil flux, the formula above must be calculated with twice the length. So there is an additional factor of 2. On this track the conductors are located in the magnetic field, just as described in Figure 27.

The identification of the velocity v is more complicated. In the end, the value of kV must be calculated. It is immediately apparent that the parameter n, the speed in rpm, is missing. Conversely as in Figure 27, where the permanent magnet is stationary and the conductor moves, in the real motor the permanent magnet moves (it is located on the rotor bell, see also Figure 15) and the conductor is fixed. However, this is indifferent. It depends only on the relative motion between the permanent magnet and conductor. The rotational speed n is calculated from the velocity as follows:

$$n = \frac{v}{\pi \cdot D} \cdot 60$$

π·D is the circumference. If v is in meters per second, the whole thing must be multiplied by 60, since n should be in rpm. Inserted this results in:

$$U = B \cdot N \cdot \frac{n \cdot \pi \cdot D}{60} \cdot 2 \cdot L$$

However, the voltage U is now only that voltage which is induced in the coil. It must now be converted to the voltage which is applied at the input of the controller. This will be called $U_{BAT}$. This is indeed the battery or accumulator voltage, and ultimately relevant for the calculation of the kV. Assuming once that there are 10V, the controller can only produce an AC voltage of half the maximum voltage, 5V. An AC voltage does pass to both the positive and negative. Furthermore, there is a delta connection in the calculation example. Thus, per 10V battery voltage 5V·√3 = 8.66V is applied at the coils. $U_{BAT}$ will therefore be calculated as

$$U_{BAT} = \frac{1}{0.866} \cdot B \cdot N \cdot \frac{n \cdot \pi \cdot D}{60} \cdot 2 \cdot L$$

And now kV is

$$kV = \frac{n}{U_{BAT}} = 0.866 \cdot \frac{60}{\pi \cdot D \cdot B \cdot N \cdot 2 \cdot L}$$

If it were a star connection, the first factor would be just 0.5 instead of 0.866. With the values determined above, this results in:

$$kV = 0.866 \cdot \frac{60}{3.1415 \cdot 0.023 \cdot 1.4 \cdot 22 \cdot 2 \cdot 0.013} \frac{\text{rpm}}{V} = 898 \frac{\text{rpm}}{V}$$

This value lies slightly lower than the 924 rpm/V stated in the datasheet. But the values of various parameters, such as the induction, were only adopted. In addition, the length and the diameter of the stator are only measured values.

### 3.2  Winding resistor

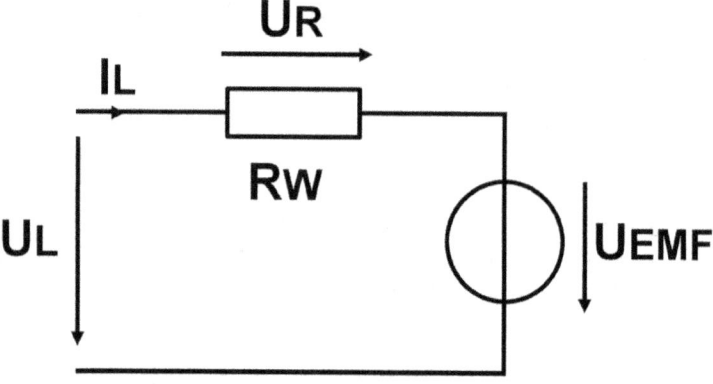

Figure 29: Simplified equivalent circuit of one of the three coils

A simplified equivalent circuit of one of the three coils will serve as an introduction to the chapter. In contrast to Figure 26, the inductance L has been omitted here. Thus, the coil will be represented for all of the following considerations as a simple resistor (just as winding resistor).

All considerations in the last chapter were made assuming that the induced counter voltage $U_{EMF}$ and the conductor voltage $U_L$ are equal. For the calculation of kV this is correct. $U_L = U_{EMF}$ is only true at idle running when the drive has no power load. Imagine this in the following thought experiment, for which a little knowledge in electrical engineering is needed: If $U_{EMF}$ is equal to $U_L$, then no voltage $U_R$ drops across the winding resistor in Figure 29. Since for each resistor Ohm's law is valid (I = U/R), then the current $I_L$ through the coil is also zero.

Of course, this hypothetical case does not correspond to practice. Indeed, friction losses exist in every real drive. As soon as such losses are present, a torque must also be applied which compensates for this. Therefore a small current always flows.

Therefore, the above hypothetical case means in reality that the idle current $I_L$ causes also a voltage $U_R$. According to Figure 29, it is also flowing through $R_W$. This is due to Kirchhoff's laws:

$$U_{EMF} = U_L - U_R$$

The induced counter voltage $U_{EMF}$ is thus smaller than $U_L$ by a voltage drop across the winding $U_R$. If one examines example 1 again with the new findings, it is found that the idle speed is not quite as high as was calculated there. The current, which is also flowing in the idle speed, ensures that $U_{EMF}$ is just slightly smaller.

This is also the reason why the speed of the drive decreases under load. One can see this simply by braking the rotor bell by hand on a drive during idle-running. The braking torque needs to be compensated with a higher current. This causes a larger voltage drop across the winding $U_R$. Because of the above formula, the effect is a smaller $U_{EMF}$ and thus a lower speed.

### Calculation of the winding resistor

To also perform calculations, it is necessary that one knows the value of the winding resistor. Most manufacturers give it on their data sheets.

The resistance of a conductor is calculated according to Figure 30 as R = ρ x length / cross-sectional area. The thicker a conductor is, the smaller the resistance is, and the longer a conductor is, the larger the resistance is. ρ is the specific resistance. For copper wire, a ρ of 17.8 E-3 Ω mm²/m is used.

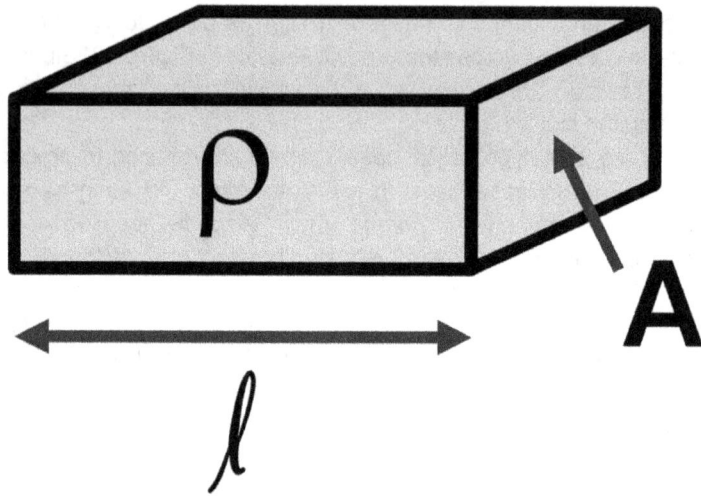

*Figure 30: Resistor with cross-sectional area A and length L*

The wire length L of a coil is calculated for a practical example to

L = 22 x (2 x 0.017m + 2 x 0.006m) + 0.1m = 1.112m

22 corresponds to the number of turns, 0.017m and 0.006m are stator length and winding-upper or lower side (also called coil end) with extra amount, because the windings are indeed on top of each other. 0.1m is an additional length, because the winding is distributed over several slots and the wire must be brought to them. The cross-sectional area A is calculated for an assumed wire diameter of 0.2mm:

A = 4 x 0.1mm² x π = 0.126mm²

It is assumed in the example that the wire is 4x in parallel due to the skin effect, so the '4' is like a multiplier. This yields a total resistance R = ρ x length / cross-sectional area = 0.157Ω. In literature, the calculations are often performed with the single-phase equivalent circuit of Figure 29. There it is always calculated from one conductor to ground (or neutral point), so all sizes should be equivalent to a star connection. If you want to wrap a wound motor in the delta connection to a star connection while maintaining the same characteristics, the number of turns and the wire length must be divided by √3. The cross-sectional area of the wire must be multiplied by √3. This is similar to the voltage ratios that were already discussed in Figure 23. Since in the resistance calculation the wire length L is in the numerator and the cross-sectional area is in the denominator, the new total resistance is smaller by √3·√3 = 3. The winding resistance $R_W$ in Figure 29 is thus 0.157Ω / 3 = 0.052Ω, if a delta connection is used in the example.

**Measurement of the winding resistor**
In practice, the coil resistance can also be measured. For this purpose some considerations are necessary. It is of course easy to connect a multimeter, adjusted to 'resistance measurement', to two of the three wires of the motor, but the results are often disappointing. The problem is that the typical resistances lie significantly below 1Ω, and in the studied example even in less than a tenth of this. Very often, the resolution of a commercial multimeter is not good enough in this range. In many cases, the measurement current will reach the limit. This distorts the results further. Those who want good results should give only a few millivolts on two of the three wires of the motor with a regulated power supply. Some caution is advised because in that resistance range the current very quickly gets too large and could possibly burn through the winding. In the present example 47.2mV at 411mA was measured, which results in a measured resistance $R_M$ of 47.2 / 411Ω = 0.114Ω. This intermediate result doesn't correspond yet to the resistance $R_W$ of the circuit in Figure 29.

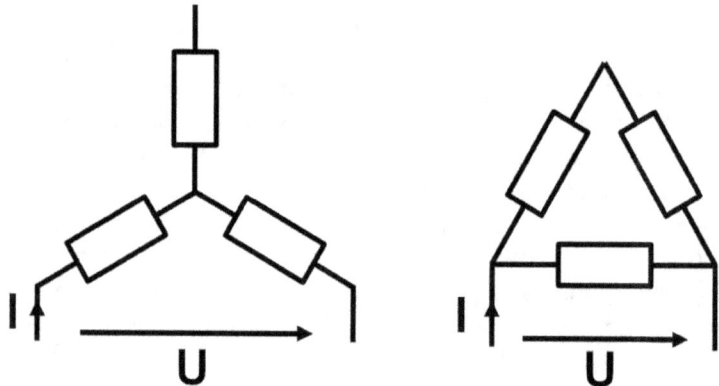

*Figure 31: Measurement of the winding resistance*

Figure 31 shows the conditions when it is measured on two wires. If a star connection is used, the relations are easy. The measured resistance $R_M$ is the series connection of two resistances $R_W$. $R_W$ is in this case equal to $R_M / 2$.

But in the example a delta connection is again adopted, as for the calculation of the winding resistance. As illustrated in the figure, the lower resistor is in parallel to the sum of the two upper resistors. Because of symmetry, all resistors are the same size. It is

$$R_M = \frac{1}{\frac{1}{R} + \frac{1}{2 \cdot R}} = \frac{2}{3} \cdot R \quad \text{or} \quad R = \frac{3}{2} \cdot R_M$$

In order to obtain the resistance $R_W$ of the circuit in Figure 29, this value must again be divided by a factor of 3 due to the conversion to star connection. Thus, here also $R_W$ is equal to $R_M / 2$, that is 47.2mV / 411mA / 2 = 0.057Ω. This value doesn't differ strongly from the calculated 0.052Ω.

A comparison of the $R_W$ between the star and the delta connection shows that it doesn't matter in the measurement what wiring exists internally. Both the winding resistance of the star connection and the winding resistance converted from delta to the star connection are always the same size as half of the measured resistance between two wires.

However, since the above considerations concerning Figure 29 describe the equivalent circuit of a single coil, the coil resistance $R_W$ is still not the resistance, which is relevant for the later calculations. Due to the special control of brushless motors, two windings are always energized, while the third remains open for measurement purposes (see Chapter 4). Relevant for the further calculations is therefore the internal resistance Ri, which represents a series connection of two identical coil resistances $R_W$. Ri is precisely the resistance which is measured between two wires of the motor, the same as $R_M$. In the above example it is measured to 0.114Ω and calculated to 2 x 0.052Ω = 0.104Ω.

### 3.3 Motor characteristics

With this knowledge, all the fundamentals are now laid to calculate and draw the so-called motor characteristics. This draws on the x-axis (the abscissa) the torque M and on the y-axis (the ordinate) the speed n.

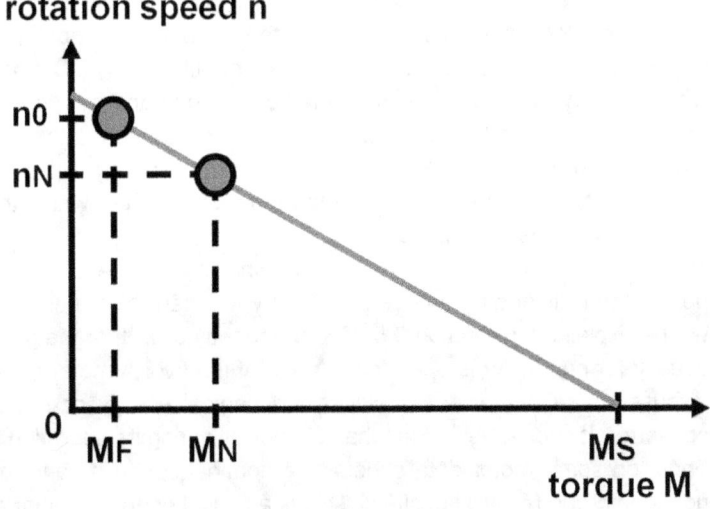

*Figure 32: Motor characteristics*

Figure 32 illustrates the correlation between these two variables. In particular, it shows that the speed is reduced with increasing torque. As will be shown later, the torque and the required current are interdependent. A higher torque also results in a larger current. This means, as shown in Section 3.2, that the voltage drop is greater across the winding resistance and therefore the induced counter voltage and thus the speed decreases. The decrease of the curve is thus a direct consequence of the internal resistance $R_i$. If the manufacturer manages to keep this small, then the curve is flatter. In the hypothetical extreme case of a superconducting coil with $R_i = 0$ it would pass horizontally. Motors with a very small $R_i$ are also known as speed-stiff. Their speed drops under load (high torque) are less strong than those of motors with large $R_i$.

$n_0$ is the idle speed. It is reached as illustrated in Figure 32, when the torque applied by the motor is equal to the frictional torque $M_F$. The idle speed of a drive could be slightly higher if it were possible to eliminate the friction altogether. At the other end of the line is the standstill torque $M_S$. It is that torque which would be at speed 0, thus at a standstill. Other terms for the standstill torque are static torque or a short circuit torque.

You may here pause for a moment and ask yourself, how can you feel a torque? Unlike the speed, it's something which can't be seen directly. For this purpose it is best to run a small (<10W) DC motor with a battery at a low speed and choke it off by hand by pushing with your thumb and index finger on the shaft. You sense that the motor would like to rotate. What you are then applying with your fingers is technically nothing more than the load torque, which the drive's torque must compensate for.

Neither in the operating point 'idle running' nor in the operating point 'standstill' does the drive deliver any mechanical power. This will be explained in Section 3.5. It is abundantly clear that the drive usually (or hopefully) is operated in none of the two. A motor is, as discussed in Chapter 1, an electro-mechanical energy or power converter. It should deliver mechanical power during its operation.

Thus, on some motor data sheets, the nominal speed $n_N$ and the nominal torque $M_N$ are specified. This is a typical operating point or even a region in which the drive delivers power at respectable high efficiency.

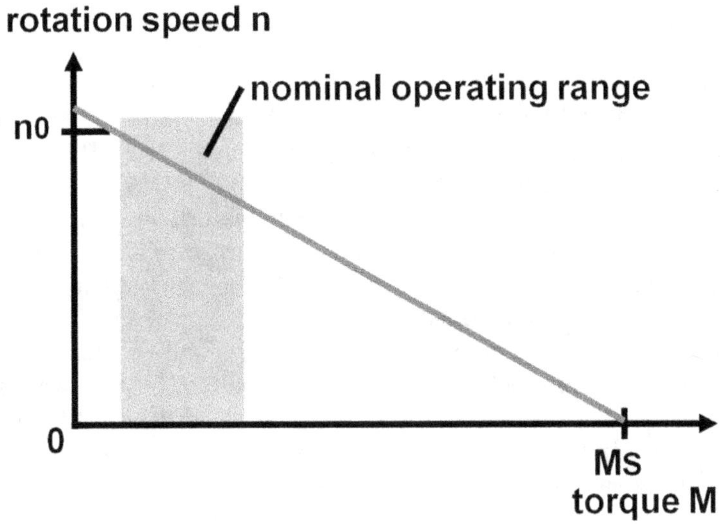

## rotation speed n

*Figure 33: Reasonable region of operation*

These two values also lie on the curve. Mostly the nominal speed is in a range between about 2/3 and 9/10 of the idle speed. Figure 33 illustrates this.

However, as a user you can't simply say: "I now operate my drive at the nominal speed". The operating points must indeed lie on the curve at any speed. There is also a corresponding torque that is demanded by the load, the propeller or the like.

**Relation between current and torque**

To a certain torque, on the electrical side flows also a specific current. Here, the DC is meant that flows between the battery and the brushless controller assumed to be lossless. These two values are related via the motor constant kM. The following applies:

$M = I \cdot kM$ and $kM = 1 / kV \cdot 30/\pi$

That the motor constant kM is inversely proportional to the value of kV derived in Section 3.1 is almost fundamental, but actually logical. If you have a motor with a large kV, i.e. a motor which makes a lot of speed per the applied voltage, this fact must be paid for with a small kM. Thus, for a certain current the motor generates only a small amount of torque. This could for example apply for an inrunner. In the opposite case, a motor which has a small kV, i.e. which makes little speed per the applied voltage, therefore has a large kM, and a specific current causes a large torque. This case could also fit for an outrunner. The factor '30/π' is only because at the end kV is calculated for a speed with the unit 'revolutions per minute'. A unit of speed without a conversion factor, a so-called SI unit, would be the so-called angular frequency. However, it is irrelevant to the considerations at this point.

Due to the proportionality between the torque and current, Figure 32 can also be supplemented by the current. Figure 34 shows the relations. For the first time the idle current $I_0$ is shown. It is that current which compensates the torque caused by the friction $M_F$. The correlations are the same as above:

$M_F = I_0 \cdot kM$  and  $kM = 1 / kV \cdot 30/π$

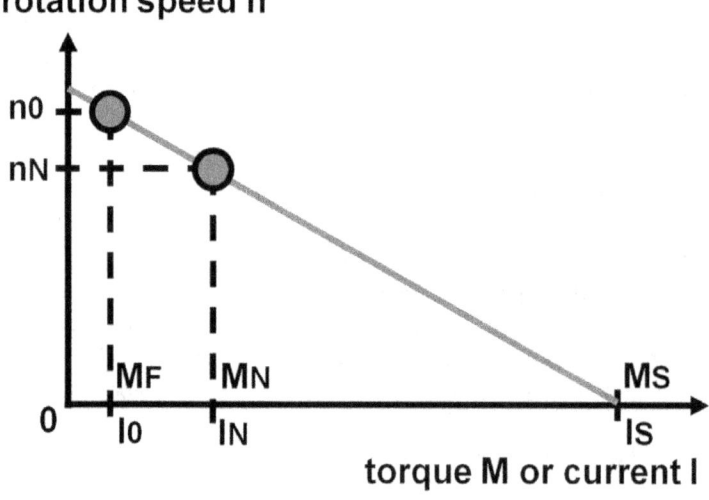

Figure 34: Motor characteristics, supplemented by the current

Furthermore, the nominal current $I_N$, which occurs at the nominal torque $M_N$ and the nominal speed $n_N$, and the short-circuit current $I_S$, which occurs at the standstill torque $M_S$, also appear.

As many have already experienced, the current is very large with a blocked motor (standstill torque $M_S$ or short-circuit current $I_S$). Then all the battery voltage drops across the typically very small internal resistance Ri = 2 x the coil resistance Rw. Figure 29 again serves as comparison. Since the motor doesn't rotate when it is blocked, the induced counter voltage $U_{EMF}$ is then equal to zero. This can also be interpreted as a short circuit, hence the name 'short circuit current'. In this case it may happen that the coil burns through. Therefore, most brushless controllers have an integrated current measurement. They can thus mostly turn off in time. This is described in more detail in Chapter 4.

## Calculation of the revolution speed at known internal resistance and current

If the current flowing from the battery and the internal resistance are known, the revolution speed n of the motor can be calculated:

$n = (U_{BAT} - I \cdot Ri) \cdot kV$

### 2$^{nd}$ example:

*A brushless motor AXI 4130/20 with the data kV = 305 rpm/V, R = 0.099Ω, $I_0$ = 1.2A is driven by a 10S LiPo. Here, a battery current of 40A (assumed to be nominal current $I_N$) is measured. The idle speed $n_0$ and the nominal seed $n_N$ are asked. In addition, the motor constant kM, the friction torque $M_F$ and the nominal torque $M_N$ should be calculated.*

*The idle speed would be at neglected idle current simply $U_{BAT} \cdot kV$ = 10·3.7V·305 rpm/V = 11'285 rpm/V, as already calculated in the 1$^{st}$ example. However, if the idle current is taken into account, one gets $n_0 = (U_{BAT} - I_0 \cdot Ri) \cdot kV = (10 \cdot 3.7V - 1.2A \cdot 0.099Ω) \cdot 305$ rpm/V = 11'248 rpm/V, so slightly smaller.*

*The nominal speed is $n_N = (U_{BAT} - I_N \cdot Ri) \cdot kV = (10 \cdot 3.7V - 40A \cdot 0.099 Ω) \cdot 305$ rpm/V = 10'077 rpm/V.*

$kM$ is $1/kV \cdot 30/\pi = 1/305 \cdot 30/\pi$ Nm/A = 0.031 Nm/A. The friction torque $M_F$ is $I_0 \cdot kM$ = 1.2A $\cdot$ 0.031 Nm/A = 0.037 Nm. The nominal torque $M_N$ is $I_N \cdot kM$ = 40 A $\cdot$ 0.031 Nm/A = 1.24 Nm.

**Comparison between measurement and calculation of the motor characteristics**

The AXI motor of the 2$^{nd}$ example was set on a drive test bench for a comparison of the motor characteristics between measurement and calculation. This motor is really quite large. It offers over 1kW mechanical power and weighs 409g.

To simulate a requested load, it was coupled with a DC motor. This generated the desired braking torque. The coupling between the two motors was also equipped with a torque meter. The above apparent friction torque is an internal matter of the AXI motor. The torque meter measures only the load torque which is applied by the DC motor. So it measures no torque at idle speed. Compared to Figure 34, Figure 35 is to be understood that the friction torque $M_F$ is moved to the origin. In this new zero point flows already the idle current $I_0$ at idle speed $n_0$.

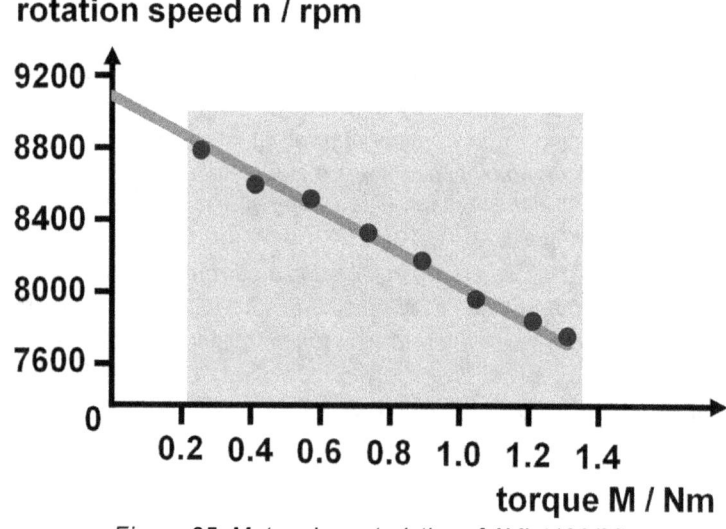

Figure 35: Motor characteristics of AXI 4130/20

To eliminate in the measurement the internal resistance of the battery and to obtain only the 'pure' motor characteristics, the brushless controller was fed by a stabilized 30V power supply, which can deliver up to 60A. The losses of the regulator were also measured and compensated at higher currents with a slightly larger input voltage. The idle speed is calculated to be $n_0$ = (30V - $I_0 \cdot R_i$) $\cdot$ kV = (30V - 1.2A $\cdot$ 0.099Ω) $\cdot$ 305 rpm/V = 9'113 rpm. The solid line represents the calculation of speed curve, whereas the dots are measured operating points. Only sensible operating points were measured (gray area). The speed drops to about 7800 rpm. As an example, the speed of the second-rearmost point is recalculated. The current I is calculated there to $I_0$ + M / kM. With a kM = 1/kV $\cdot$ 30/π = 1/305 $\cdot$ 30/π Nm/A = 0.031 Nm/A is this 1.2A + 1.2/0.031 A = 39.53 A. The speed n is then (30V - I $\cdot$ $R_i$)$\cdot$ kV = (30V – 39.52A $\cdot$ 0.099Ω) $\cdot$ 305 rpm/V = 7'956 rpm.

**Throttle position and motor characteristics**

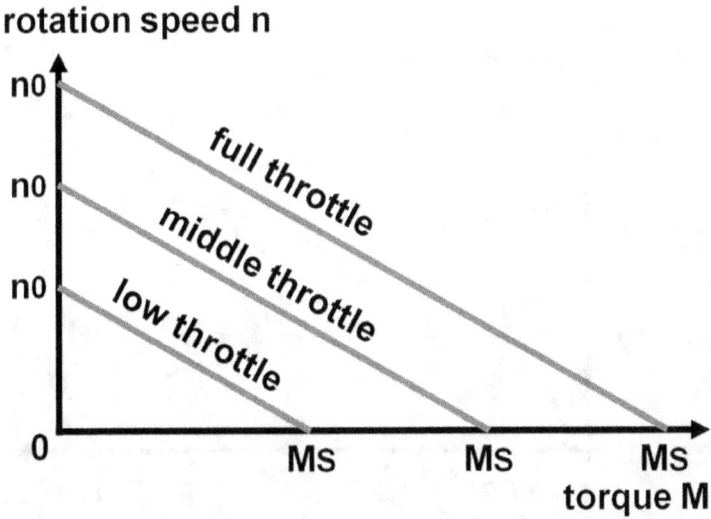

*Figure 36: Motor characteristics at different throttle positions*

There is the further question, what the influence is of the throttle position on the curve. Here the next chapter is already somewhat anticipated. There the brushless controllers are discussed in detail. Simply described, the position of the throttle means that the controller doesn't convert the full battery voltage to an AC voltage, but only a percentage corresponding to the throttle position. Because the kV of the motor is still the same, this causes a lower idle speed. As explained above, the speed drop at higher currents and torques is due to the winding resistance. Since this always remains the same, the curves run parallel to each other as shown in Figure 36.
As can be seen, this also changes the standstill torques $M_S$. Even that can be easily understood in practice. It is at a low throttle position with a smaller idle speed much easier to stall a small few Watt drive (<10W) by hand, since one must then apply a smaller standstill torque $M_S$.

**Operating point**
But the motor is not running on the whole characteristics, but only at an operating point, in the technique also called working point. The load, i.e. what it drives, determines it.

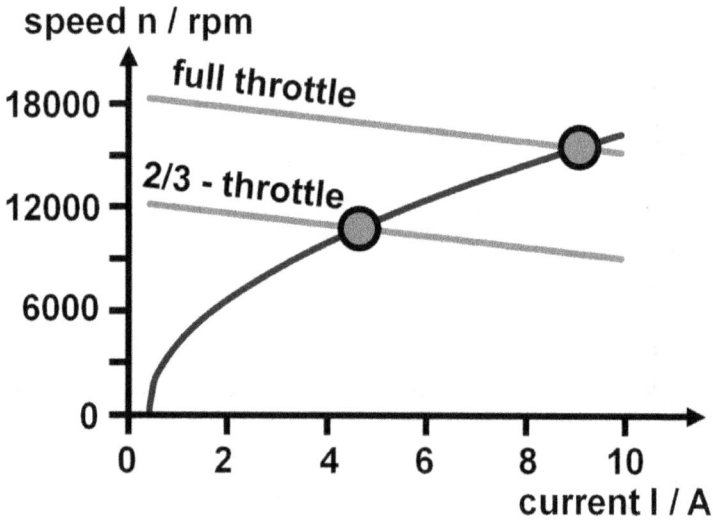

Figure 37: Scorpion SII2205-1585 and APC E 6" x 5.5"

With a model airplane, the load is the propeller. As a practical example this shall be shown with a Scorpion SII2205-1585 motor and an APC E 6" x 5.5" propeller. Figure 37 shows the relations.

The two motor characteristics at full throttle position and throttle position to 2/3 gas are shown in the figure as straight lines. The propeller has a so-called non-linear characteristics. It is also called load line. At low speeds, the current (and thus the torque) is very small, while at high speeds it increases disproportionately. Basically, all boat and airplane propellers show these properties. One may wonder why the characteristics of the propeller doesn't begin at the origin (0 A at 0 rpm). The reason for this is again to be found with the idle current of the motor.

The operating point is now exactly where the motor and load characteristics intersect. A 3S LiPo accumulator with 11.1V was assumed. The operating point is read off at full throttle at about 14,500 rpm and 9.5A. If the throttle position is set to 2/3, only the motor characteristics changes. The propeller remains the same. Thus, the new operating point simply moves to the new intersection of the two lines. The new values are then approximately 10,500 rpm and 5A.

## Drop in speed at high torque in practice

In internet forums, many measurements of motor and propeller combinations are found. If you browse around a bit in these and compare the characteristics, then it is often the case that the speeds drop more strongly with increasing currents and torques than was calculated in the examples above.

This is also correct in practice. The internal resistance of the brushless controller and the battery to the internal resistance of the motor are then added, and the supply lines are also not without resistance. If one wants to realize a drive as speed-stiff as possible, one must always make sure that the whole chain of battery pack over the controller to the motor has a small internal resistance. In addition, the supply lines should always be kept as short as possible.

## 3.4 Turns

In many data sheets 'turns' is specified as one of the characteristic sizes. For a discussion of what it's all about, two previously derived formulas are repeated once again:

$M = I \cdot kM$ and $kM = 1/kV \cdot 30/\pi$
and
kV is proportional to $1 / (\pi \cdot D \cdot B \cdot N \cdot L)$

A motor with a large kV, which makes many rotations per applied volt, therefore has a small kM because of the inversely proportional relation. It has very little torque per ampere, or in other words, if the load requires a certain torque, it requires a high current. 'Turns' is nothing other than another term for the number of turns per coil, N. N stands for the calculation of kV in the denominator and for the calculation of the kM in the numerator. An investigation of the formulas shows:

*Many 'turns' means a small kV and a large kM. The motor rotates slowly and for a given torque, it needs a small current.*
*Few 'turns' means a large kV and a small kM. The motor rotates quickly and for a given torque, it needs a high current.*

Many motor manufacturers will vary the number of turns in their model series. Motors with fewer turns are colloquially referred to as 'evil' motors. They rotate faster, for example, make an RC car faster and also stress the brushless controller more, because it has to deliver a larger current for the required torque. In addition to the maximum current, how many 'turns' a motor should have for operation is also mentioned in the data sheet of many brushless controllers.

## 3.5 Motor power

Caution is required for the term 'power'. It is important to define what kind of power is meant exactly: is it the electrical power, the mechanical power or the power losses? As mentioned at the very beginning of this book, a brushless motor is nothing other than an energy converter, which converts electrical energy into mechanical energy with more or less losses. However, one can also speak of the term power, since the energy is nothing other than the time-integrated power. If the power is constant, as is the case with most observations in this book, then the words in the last sentence are simplified by 'energy = power x time'.

Thus, the brushless motor can also be regarded as a power converter, which converts electrical power into mechanical power. Figure 38 shows how this is to be understood.

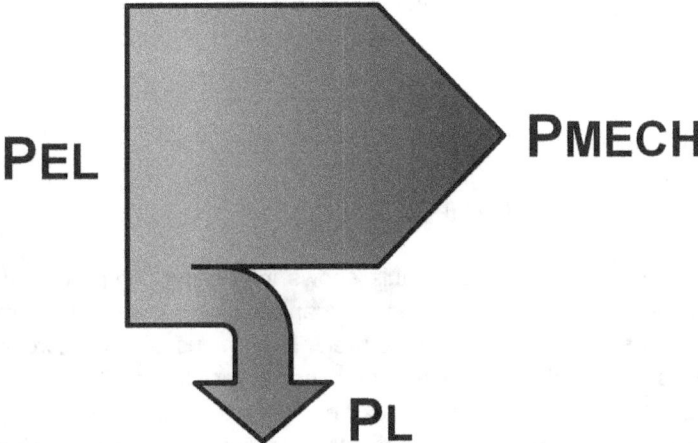

*Figure 38: Electrical power, mechanical power and power losses*

As can be seen and already shown at the beginning of the book, here is:

$P_{EL} = P_{MECH} + P_V = U_{BAT} \cdot I$

This is actually related to the input of the brushless controller. This, however, is assumed in the following considerations as lossless. The following equations now describe the relation between the electrical and mechanical parameters:

$n = (U_{BAT} - I \cdot Ri) \cdot kV$

as already previously used and

$M = kM \cdot (I - I0) = 1/kV \cdot 30/\pi \cdot (I - I_0)$

Here, on the one hand, the relation between kM and kV is considered. On the other hand, the idle current is anchored in the formula. It contributes nothing to the torque and must be subtracted from the current I. In the explanations in Section 3.3 this has been taken into account; the current $I_0$ and the friction torque $M_F$ have been moved in the zero point of the motor characteristics. The mechanical power at the shaft is calculated to

$P_{MECH} = n \cdot M \cdot \pi/30$

If the above formulas with n and M are used, this results in:

$P_{MECH} = (U_{BAT} - I \cdot Ri) \cdot (I - I_0)$

The discussion of this formula shows interesting things. It would actually be desirable that the mechanical power is if possible the same size as the electric power. Two minus signs, however, worsen this desire. There are winding resistance Ri and idle current $I_0$, which characterize the subtracted terms. The motor manufacturers optimize their motors by trying to keep these values as small as possible.

### 3rd example:
*A brushless motor AXI 4130/20 with the data kV = 305 rpm/V, Ri = 0.099Ω, $I_0$ = 1.2A is driven by a 10S LiPo. Here, a battery current of 40A is measured. Using this, the electrical and mechanical power should be calculated.*

*$P_{EL}$ is calculated to $U_{BAT} \cdot I = 10 \cdot 3.7V \cdot 40A = 1480W$.*

*$P_{MECH}$ is $(U_{BAT} - I \cdot Ri) \cdot (I - I_0) = (10 \cdot 3.7V - 40A \cdot 0099Ω) \cdot (40A - 1.2A) = 1282W$.*

The difference in power, the 1480W - 1282W = 198W, is thus the power loss $P_L$ of the motor. It must be dissipated in some way. With model airplanes, it is often sufficient to place the motor in the propeller air stream, or at least to ensure with the drive cover that a little residual air flows to the motor. Often you see in car models cooling ribs directly on the motor circumference. There, the weight issue is not so critical.

Contrary to what you might know from earlier, when the brushed drives were sometimes almost boiling hot due to the very large power dissipation, this is not the case with today's brushless motors. Sensible designs and compliance with the specifications of the manufacturer, especially of the maximum current, ensure that they will only become lukewarm or slightly warmer in operation. The small loss of power of the drive makes this possible and the motor shows its appreciation with a long lifespan.

**Relation of size and power**

*Figure 39: Cylindrical surface A of a stator*

With a larger motor also more power is achieved. This correlation is obvious and is hardly questioned by anyone. In this section it will be investigated more closely. In various sources, the power is given proportional to the stator surface. As a rule of thumb, the following formula can apply for a motor without a built-in fan:

$P_{EL} = (0.05...0.1) \text{ W/mm}^2 \times A$

As Figure 39 shows, A is the cylindrical surface of the stator. A balance of motor temperature adjusts itself when the dissipated heat is the same size as the one resulting from the power loss in the motor. Heat can be dissipated in several ways.

- a.) Convection: cold air flows over a hot surface; it heats up and thus provides a cooling of the surface. A PC fan is a good example.
- b.) Heat conduction: a hot surface is connected with a good heat conductor, so this also heats and cools the surface this way, because the heat energy will be taken away from it. An example is a metal pan on an electric stove.
- c.) Heat radiation: a hot surface also cools down if it is surrounded by cold objects. Solar radiation is an example of this. A temperature sensor which is exposed to direct sunlight shows a higher temperature than one that measures in the shade, so in a cooler environment.

All three kinds share the characteristics that the size of the surface area is relevant for heat dissipation. The factor from $0.05 \text{W/mm}^2$ up to $0.1 \text{W/mm}^2$ is based on experience. Depending on mounting, a smaller or larger power is possible; therefore a range is given. If the drive is in the air stream of the propeller, the convection is greater than if it must turn in a housing without air circulation. When the drive sits on an aluminum carrier, the heat conduction is also better than when it is mounted directly on a carrier made of wood. If the drive is also equipped with a fan, this factor may even be more than $0.1 \text{ W/mm}^2$.

Until now, however, we have only considered the power loss and not the electrical or mechanical power itself. If we consider Figure

38 enlarged or reduced in size, all powers will be enlarged with a larger picture and smaller with a reduced picture. So they always depend on each other. Thus the above rule of thumb will be explained.

### 4$^{th}$ example:

*For a motor Roxxy 2827-34, the allowed electrical power can be calculated according to the rule of thumb above. The stator diameter D is measured as 25 mm and the length L is 15 mm. The cylindrical surface of the stator is calculated as $2 \times (D/2)^2 \times \pi + D \times \pi \times L = (2 \times 156.25 \times 3.14 + 25 \times 3.14 \times 15)$ $mm^2$ = $2159 mm^2$. $P_{EL}$ is calculated with the conservative factor of 0.05 $W/mm^2$ to 108 W.*

The motors calculated in this way can be operated in any case with a continuous power of this size. Often, the manufacturers give a higher maximum power in their data sheets. It might be up to twice as high. They in addition specify for what time period (e.g. 15 seconds) this is possible. During this time the motor warms up strongly. But if it is then turned off for a long time, it can cool down again. This is especially interesting for models of motor gliders.

## 3.6  Braking operation

All of the above explanations are only true when the drive is running in the so-called motor operation, so if it makes mechanical power at the shaft. For the drives which are used in model airplanes and boats, that's the normal case. For car models, however, it is not only the acceleration that is important. They also need to slow down as efficiently as possible when approaching bends. Also with models of electric gliders, the braking operation is important, so that the flip airscrew can flip back in a well-defined manner. The technical term for this is called the generator mode. Often one hears in this context the term 'EMF-brake'.
The word 'generator' is actually familiar with the generation of electricity. Such generators are used in power plants. But the generation of electricity is exactly the case in the braking mode of the operation. Figure 40 shows the overview of powers.

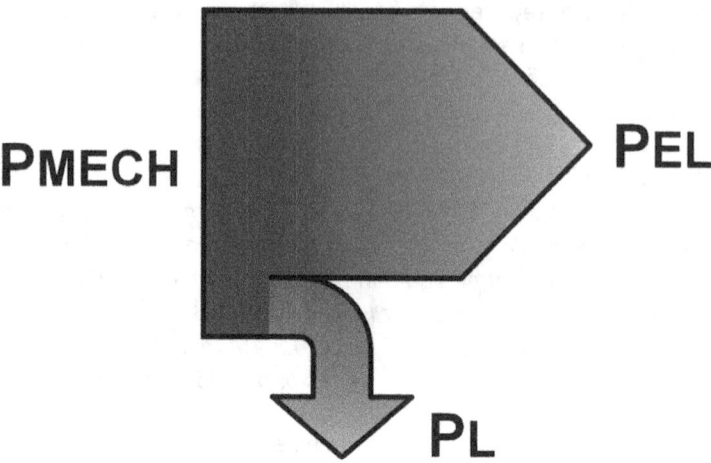

*Figure 40: Powers in brake or generator operation*

In contrast to Figure 38, now mechanical energy is converted into electrical energy or just mechanical power into electrical power. This isn't lossless either. It also produces power losses in the generator (or in the motor, which is operated as a generator).
However, it is the same machine, which depending on the operating case makes the conversion electric power → mechanical power (motor) or mechanical power → electrical power (generator). The following example will serve as an explanation.
It is assumed that a combination consisting of accumulator, brushless controller and a brushless motor at slow speed is driving a hamster wheel. For simplicity, everything is lossless; in particular there is no idle current $I_0$. If the hamster turns its rounds in the wheel as fast as the motor rotates, then this is idle-running. If the hamster wants to run faster, it tries to accelerate. The hamster now powers the motor, which is now rotating slightly faster and is now called the generator. The energy flow is reversed.
Figure 41 shows the effect. The line of all operating points is simply drawn over the idle speed. Since the motor doesn't drive the hamster wheel, but the hamster wheel drives the generator, it results in a negative torque. In the figure it is shown as a braking torque $M_B$. As explained above, current and torque are related to each other. Thus, the current also becomes negative. Finally, a

negative current is flowing from accumulator to the brushless controller, or in other words, a positive current is flowing from the brushless controller to the accumulator. Thus, the accumulator is recharged.

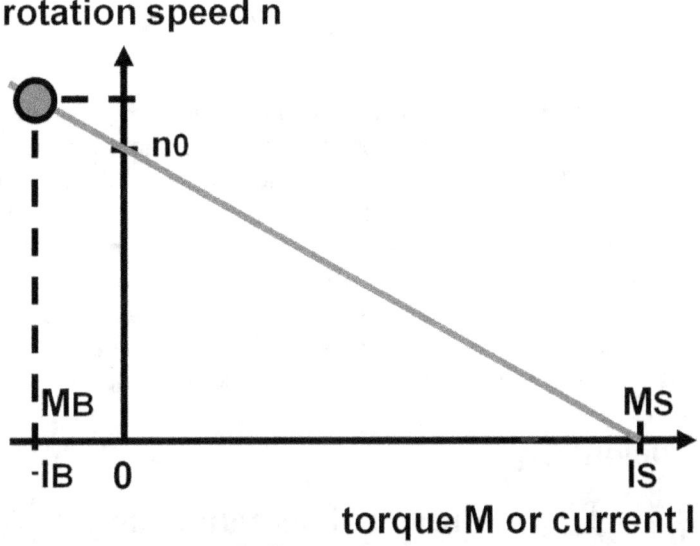

*Figure 41: Motor characteristics for braking or generator operation*

In practice, the brushless controller moves back in the braking operation simply more or less quickly (hard or soft braking) from full throttle to less throttle by the curves of Figure 36. This reduces the required speed more or less quickly. The model car or the propeller is braked more or less rapidly and the energy is fed back into the accumulator. With a modern brushless controller, the braking energy is not converted into heat, but it charges the accumulator. This extends the operating time of the drive.

### 3.7 Efficiency

There is now only a small step to determine the last of the important characteristic of a brushless motor, namely the efficiency. The efficiency of a motor is calculated as the ratio between the mechanical and electrical power. It is, therefore, taking into account the above formulas:

$$\eta = \frac{P_{MECH}}{P_{EL}} = \frac{(U_{BAT} - I \cdot R_i) \cdot (I - I_0)}{U_{BAT} \cdot I}$$

If superconductivity were available and no friction torque present, so if Ri and $I_0$ were zero, then the numerator and denominator would be of the same size and the efficiency η would be in any operating case at 100%. In practice, it is the case that η is zero for the idle-running. Then I = $I_0$ and the right term in the numerator is also zero. This can be explained simply because the mechanical power $P_{MECH}$ = n · M · π/30 is also zero in this case. The idle-running is so defined that there applies no torque M. But, nevertheless, the drive then needs the electric power $U_{BAT}$ · $I_0$.

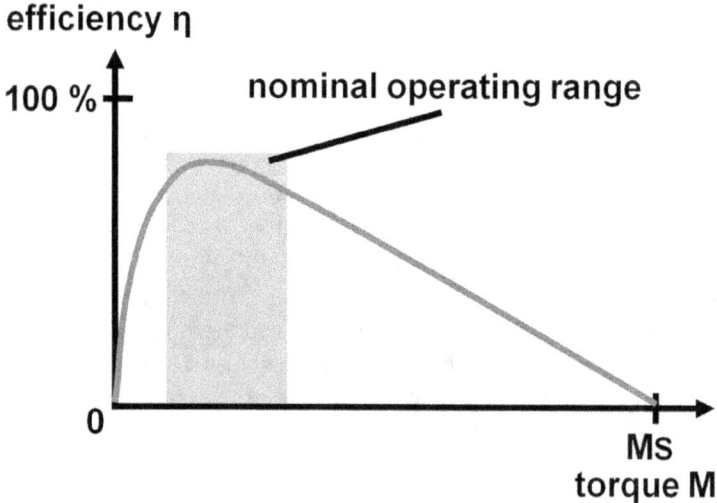

*Figure 42: Efficiency*

At the other extreme of the standstill is Ri = I · $U_{BAT}$. This case is only observed with blocked wheels or propellers. Here the left term in the numerator would be zero. Again, the drive doesn't generate any mechanical power and the efficiency is zero. The sensible operating range of a drive is illustrated in Figure 42. As a practical example, the efficiency of a Kontronik Tango 45-06 is investigated. This motor is a special version of an inrunner with a so-called air gap winding and iron lossless technology. A schematic diagram is shown in Figure 43.

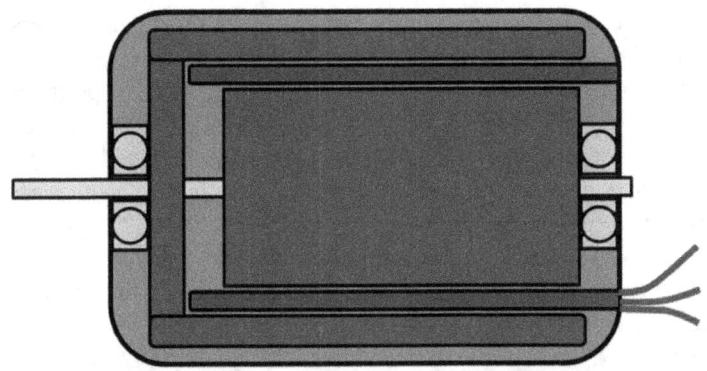

Figure 43: Motor with iron lossless technology

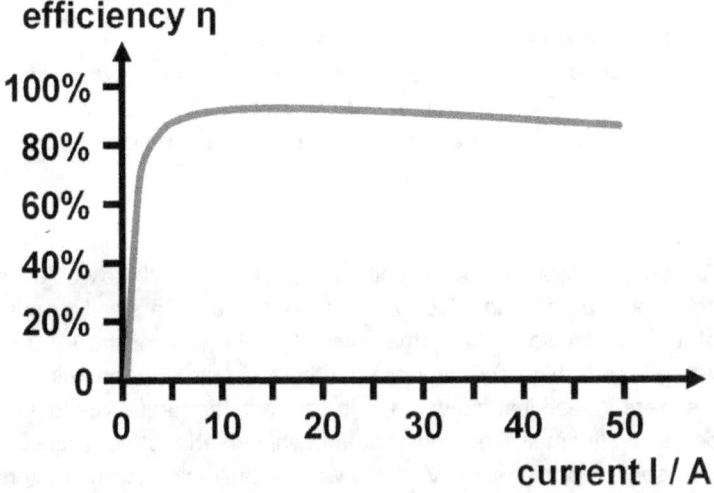

Figure 44: Calculated efficiency with iron lossless technology

Such motors were already discussed in Chapter 2 (Figure 13). As discussed there, they have no cogging due to the lack of slots. Therefore the idle current remains small and the efficiency is high. This motor forms a specialty with the iron lossless technology. Normally with an air gap winding, both the winding and the casing of iron belong to the stator, so they are not movable. When the permanent magnet rotor rotates, the iron of the housing will always be magnetically reversed. This becomes noticeable as an additional loss with a larger $I_0$. But in a motor with iron lossless technology, the iron rotates itself together with the rotor and its permanent magnets. Thus, it will not be magnetically reversed. In Figure 43 permanent magnets and co-rotating iron are illustrated. They are both part of the rotor and mechanically connected to the shaft. Technically the production is not easy to realize. Consider that there is a 'free-flying winding' between two rotating parts, the permanent magnets of the rotor and the iron for the magnetic backflow. In the picture it is drawn in between of the rotor. The achievable efficiency is remarkable. Figure 44 shows the representation of a calculation. It is located across the specified operating range from 10A to 40A around 90% and for currents between 10A and 20A reaches dream values of up to 92%.

**Further measures to increase the efficiency**
As mentioned above, the internal resistance Ri and the idle current $I_0$ are relevant for the efficiency. It will be described below to what the manufacturers pay attention to keep these two sizes as small as possible.

**Small Ri:**
Ri depends directly on the coil resistance. A small Ri would be achieved if a short and thick wire were wrapped in a small number of turns N. However, after the calculation formula for the kV, this would lead to high speeds because there is N in the denominator; there are indeed just a few turns. In contrast, the torque would then be relatively small. The constant kM, which leads to the torque, is inversely proportional to kV. If you want to build high-torque motors for direct drives, you have no other choice than to wind up thinner wires with a larger number of turns. That's bad for the Ri in two

ways. A greater number of turns leads to a greater length of wire and the thin wire to a smaller cross-section area. Both increase the coil resistance. Especially for outrunners, with which we would also reach a large torque, there is still something to consider. We browse again back to Figure 28. It is of course only the vertical length L where the conductor contributes to the force generation. The winding heads on the bottom and top do not produce any forces which can rotate the rotor. As such they are useless. If one compares the stator shape of outrunners with inrunners, the outrunners are typically short and thick, while the inrunners are long and thin. Long and thin inrunners have proportionately less 'useless' copper content than the short and thick outrunner, whose windings must be turned after a short distance. Torque is equal to force x radius. At larger radii there is more hollow space for the copper coil available. Therefore, one can often realize the same number of turns with wires of somewhat larger cross-sectional areas. This reduces the Ri again.

If one examines again the formula for kV, which is indeed inversely proportional to the kM, it is visible that the induction B as well as the number of turns N is in the denominator. A low speed or a large torque can thus be achieved even by increasing the induction B. The neodymium magnets frequently used today achieve in combination with small air gaps dream values of up to 1.4 Tesla. Also thanks to this material, you can today realize a smaller number of turns with wires of a larger cross-section area and thus a smaller Ri.

**Small $I_0$:**
The idle current is reduced strongly in comparison to the brushed motors because the mechanical contact with the shaft is now only via the ball bearings. However, besides the mechanical bearing losses there are also others which affect the friction torque and thus the idle current.

With normal inrunners, the iron for the backflow doesn't rotate; it is identical to the housing (for comparison see Figures 11, 12 and 13). Iron is also a good conductor. An alternating magnetic field acts exactly equal to conductive materials, such as when they are moved in a field (Figure 27). Thus a voltage is induced locally and therefore local currents, so-called eddy currents, also flow, leading to a warming of the iron. This heating also contributes to the loss of power (x time) and must be supplied by the battery. It leads therefore to a higher idle current. It is greater the faster the idle speed is, because then the magnetic field is also changing faster.

That's the case even with outrunners. Here the magnetic flux runs back through the inner stator. For this reason, the stators are made with a sequence of metal sheets. They are insulated from each other by a lamination. Thus although a voltage is still induced, only a small current can flow. The stringing together of all sheets of metal is then the so-called lamination stack. In practice, the iron sheets are usually made in a thickness from about 0.2mm (high quality and expensive motors) to 0.5mm (low cost) motors. A larger thickness has almost no effect any more, while at a smaller thickness the effective length of iron is reduced too strongly, due to the lamination.

For the same reason, the permanent magnets of the rotor in inrunners are often built in mutually insulated discs.

The cogging torques considered above also produce additional eddy current losses. However, they are much smaller than those which are caused by reversing the polarity of the magnetic field.

Finally, also the so-called hysteresis losses should be mentioned. In iron there are dipoles, which align themselves to the magnetic field. They are ultimately responsible for the magnetic effect. As a small experiment you can magnetize a nail with a permanent magnet. When this is removed, it is still slightly magnetic because the dipoles remain in their position. The reversing of the polarity of the magnetic field forces the dipoles to also change their direction. It will also generate some heat and this work must also be done by the battery.

Mechanical losses, eddy current losses due to reversal of magnetization, eddy current losses due to the cogging torque and hysteresis losses all contribute to a larger idle current $I_0$.

## 3.8  Findings in brief

Also in this chapter the main findings are summarized again.

The characteristic kV or rpm/V is the most important one of a brushless motor. If it is multiplied by the accumulator voltage at the input of the brushless controller, the result is about the maximum idle speed of the motor.

The characteristic size kM is inversely proportional to kV. It describes the relation between the torque and the motor current and is calculated as $kM = 1/kV \cdot 30/\pi$.

The smaller the winding resistance is, the stiffer the motor speed is and the less the speed drop under load is. In practice, the speed drops with increasing motor power (and torque) mostly more strongly than calculated, because to the winding resistance must be added the resistance of the wires and the internal resistances of the battery and brushless controller.

The optimum range of operating speed is approximately between 2/3 and 9/10 of the idle speed. The efficiency decreases when operating close to the idle speed or below 2/3 of the idle speed.

Many 'turns' means a small kV and a large kM. The motor rotates slowly and for a given torque, it needs a small current.
Few 'turns' means a large kV and a small kM. The motor rotates quickly and for a given torque, it needs a high current.

Regarding the term 'power' it is important to distinguish between the electrical and the mechanical power. In a motor, the power loss is the difference between electrical and mechanical power. The power loss is responsible for the warming of the motor.

In the braking mode the motor operates as a generator or as an EMF-brake. Both power and torque are negative. In this case, energy is fed back into the accumulator where it will be recharged.

*The motor efficiency is the ratio between mechanical power and electrical power. The idle current and the coil resistance worsen it. Therefore the manufacturers try to keep these two sizes as small as possible.*

## 4. Brushless DC Controllers

With the functionality and characteristics of brushless motors now explained, this chapter is dedicated entirely to brushless controllers. As discussed in Chapter 1, brushless motor and controller always form a unit. In this chapter the basic operations will be discussed with some examples. Chapter 5, in which typical practical examples of controllers and motors are discussed, will often refer to Chapter 4.

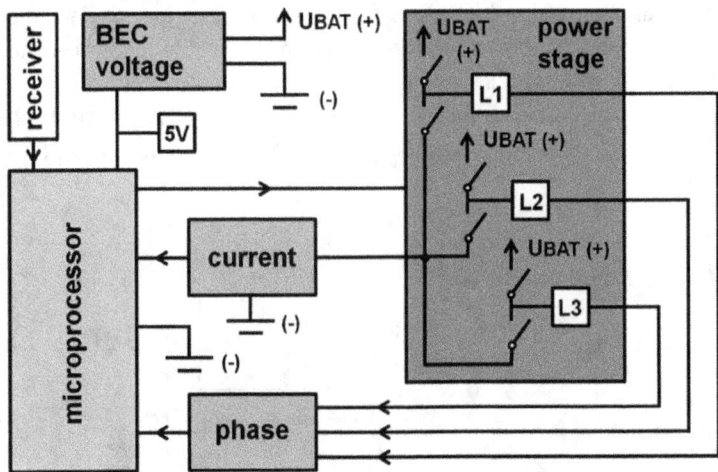

*Figure 45: Brushless controller with back-EMF measurement*

In Figure 45, all components of a modern brushless controller are illustrated. The function of the individual blocks and their interconnections with each other will be discussed here only in overview. A separate sub-chapter is then dedicated to each block.

Two very large blocks can be seen immediately, the **microcontroller** and the **power stage**. They are the main components of a brushless controller. The power stage is at first presented as a combination of (electronic) switches. The three conductors L1, L2 and L3 leading to the brushless motor are always positioned in the middle of two switches. A total of six of these exist. Each one is driven by the microcontroller. Thus, six data lines lead to the power stage (for clarity only one is shown).

The switches ensure that at the three conductors only the positive battery voltage $U_{BAT}$ (+), negative battery voltage (GND, -) or no voltage is applied. The correct sequence of fast positive and negative battery voltage generates the desired sinusoidal voltage and current.

The microcontroller must be equipped with information about the rotor position. Therefore, a **phase measurement** is performed. Many features and advantages of the brushless motor and controller are based on this.

The **current measurement** and **BEC voltage** do not belong to the core function of the brushless controller, but they are usually also part of this. With the current measurement, the controller has additional information about the motor current. With the BEC voltage other consumers in the system are supplied from the battery. In model construction, these are the receiver and the servos.

**4.1    Power stage**

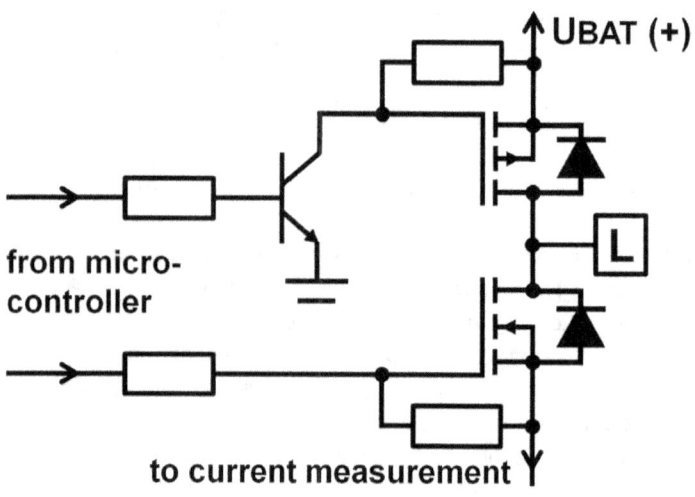

*Figure 46: Electrical circuit for the control of two switches*

Figure 46 shows the circuit diagram for one of the three conductors. The whole output stage consists of three identical such circuits. The switches are realized in the electronics by so-called MOSFETs (metal–oxide–semiconductor field-effect transistors). The lower one of the two is a so-called N-channel MOSFET.

If the microcontroller has 5V (logical 1) at the output, the MOSFET switches on, which means that it connects the output L with ground, i.e. with the negative battery voltage. If the microcontroller has 0V (logical 0) at the output, it blocks, which means that Ground and L are disconnected. At the upper MOSFET it is slightly more complicated because of the matter of the voltage level. If the microcontroller output is 5V, first the preceding bipolar transistor is activated. At the input of the MOSFET, known as gate, 0V is then applied. Since it is a so-called P-channel MOSFET, it switches on also in this case. It then connects the positive battery voltage with the output L. If the microcontroller output is 0V, the MOSFET blocks. The positive voltage of the battery is then separated from L. Altogether, three switching combinations are possible:

| Upper MOSFET(P) | Lower MOSFET(N) | Output L |
|---|---|---|
| blocks | switches on | negative battery voltage |
| switches on | blocks | positive battery voltage |
| blocks | blocks | no defined voltage |

A fourth combination, in which both MOSFETs switch on, must be avoided by the microcontroller under all circumstances, since then the battery voltage would be short-circuited. In this case, at least one of the two transistors would burn through immediately due to the high current. There are also other ways to realize the power stage. Because the N-channel MOSFET is superior to the P-channel MOSFET in terms of its conductivity, often also the upper MOSFET is realized as N-channel. For the control special driver stages are then used.

**Trapezoidal voltage**

As has been discussed in Chapter 2, the brushless motor requires a three-phase sinusoidal current. Using brushless controllers with $U_{EMF}$ measurement (see Section 4.2), this is achieved only approximately, because the voltage of the conductors L is trapezoidal in shape. Figure 47 shows the voltage curve at one of the three conductors against the negative battery voltage. The question may now arise of how the current behaves with such a voltage. Each motor coil includes in addition to the winding resistance also an inductance. This was already shown with the single-phase equivalent circuit in Figure 26, but not yet considered. An inductance causes a delay of the current against the voltage curve. With a trapezoidal voltage signal, a smoothing of the corners is the consequence. Thus, thanks to the inductance, the current curve is much more similar to the desired sine wave than the voltage. In good approximation we may therefore also speak confidently of a sinusoidal current and the discussion in Chapter 2 is still valid.

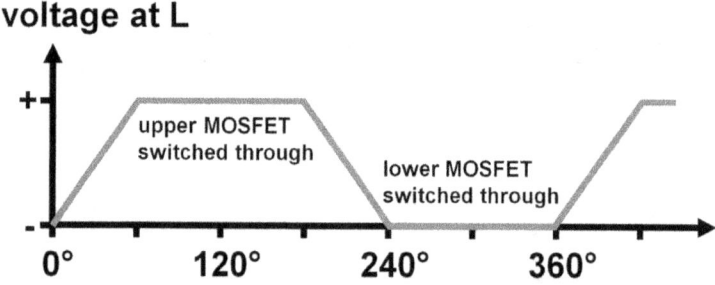

*Figure 47: Trapezoidal voltage curve at one conductor*

A complete run of a period can also be specified with an angle of 360 degrees instead of the time. During 120 degrees the upper MOSFET is switched through and the conductor L has the positive battery voltage. During another 120 degrees, the lower MOSFET is turned on and the conductor L has the Ground of the battery, while 2 x 60 degrees block both MOSFETs and the voltage rises or falls.

### Part load behavior, pulse width modulation

The above case applies only to the maximum speed. Due to the kV, the number of revolutions per minute of a brushless motor and the battery voltage are coupled to each other. Many applications require an operation at partial load, so not with the full speed. In model construction this is the case with helicopters, boats, cars, quadrocopters and often airplane models. When the motor is running at a lower speed, the full battery voltage cannot be applied. In practice, however, because of the MOSFETs, either plus or minus battery voltage is applied to the conductor; there is no 'in between'. The inductance helps here too. If the voltage is switched fast enough between the two states, then the current can't react fast enough because of its delay, and the same behavior is achieved as if something 'in between' were applied.

The following thought experiment will serve for a better understanding: If you switch a light on and off very quickly, then the eye is no longer able to recognize this due to its slow reaction. A TV also works In this way. More than 20 frames per second appear as a film, or as the reality. If the time during which the light is turned off now lasts longer than the time during which the light is turned on, then it will appear darker, and in the reverse case it will appear brighter.

In technical terms this is also called pulse width modulation or PWM for short. Figure 48 shows the effect of pulse width modulation on the voltage of the conductor.

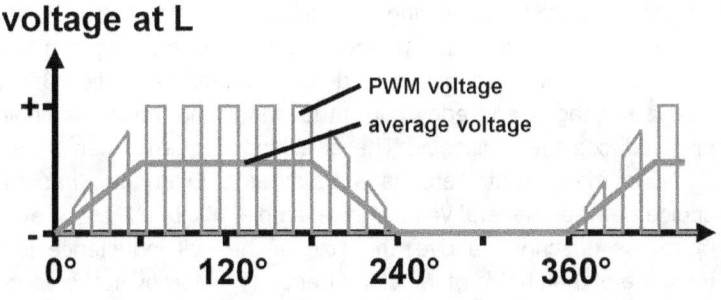

*Figure 48: Trapezoidal voltage curve at one conductor with PWM*

When the angle is from 60 degrees to 180 degrees, the upper transistor is no longer only switched on, but the lower and upper alternate. Relevant for the motor current is then just the average voltage. In the shown case, when both transistors are switched off and on for the same length of time, the average is just half of the battery voltage. If the upper MOSFET is switched on for a longer time and the lower one for a shorter time, then the average would be larger. In the opposite case it would be smaller.

At partial load, special attention must be given to the heat in the regulator. When it is driven with 'active freewheeling', always one of the two transistors is switched through. Thus it pulls the motor connector with low resistance to the negative or positive battery voltage. This keeps the power dissipation in the brushless controller small. The manufacturers then also note in the data sheets that their regulators are fully capable of partial load.

**Switching frequency**

The frequency with which the pulse width modulation switches back and forth is the switching frequency of the brushless controller. The larger it is, the less it affects the current curve of the motor, because of the inductance. However, each switching operation generates losses in the controller, the so-called switching losses. One can imagine this that during switching neither the current nor the voltage in the switch is zero. Therefore a loss of power (= voltage x current) results. It passes roughly proportionally to the switching frequency. Moreover, some losses also occur in the motor due to the switching frequency. Since the current slightly varies despite its delay by the inductance, also the magnetic field varies slightly. The variations lead as described in Section 3.7 to induced voltages and eddy current losses in the iron. This finally also reduces the efficiency. The switching frequency can be set today in many controllers as a parameter. Often one has the choice between several values, for example 16kHz, 32kHz or even more. As described above, the size of the coil inductance is a measure of the choice of switching frequency. Moreover, the speed (and the number of poles) is important, because the sinus frequency of the current must be correspondingly higher at higher motor speeds and thus the switching frequency must be even higher. It is something like the following:

*Outrunners with low speed typically have many turns and a large motor inductance. The switching frequency should be set low, for example below 16 kHz.*
*Outrunners or inrunners with a high rotational speed typically have fewer turns and a medium-sized motor inductance. The switching frequency has to be set medium-high, for example to 16 kHz.*
*Inrunners with ironless coil and a high speed typically have only a few turns and a small motor inductance. The choice of switching frequency has to be set high, for example 32 kHz or more.*
*When in doubt, the lower switching frequency should be set, because of the switching losses.*

**Practical example**
Figure 49 shows the voltage curve on a conductor for the motor and controller combination Hacker A30-16M and X-40-SB-Pro. The operation is idle-running.
A comparison with Figure 48 shows that the trapezoidal voltage is reversed. There, the lower MOSFET turns fully through while 120 degrees. Here, however, the upper MOSFET turns fully through while 120 degrees. The two modes of operation are equivalent. The motor current is the same whether the voltage is switched between 0V and 4V, or between 6V and 10V.
Since there is an outrunner motor with a relatively large inductance, the switching frequency was set to 8 kHz. Counting the switchings between the upper and lower MOSFETs, for example between 0.5 ms and 1.5 ms, we get eight per second; the result is 8 · 1000 = 8000 changes.

In the measurement one also sees the average voltage. This is so because, in practice, the lower transistor is not switched off and the upper transistor is not switched on at the same time. Because of the switching time, it could happen that for a brief moment both the upper and the lower transistors switch through and there is therefore a short circuit between positive and negative poles of the battery via the two transistors. A brief moment is therefore always required in between during which both transistors are turned off. Then the induced voltage of the motor is at the conductor. It corresponds exactly to the average according to Figure 48, because the characteristic value kV combines average voltage and speed. Therefore, the trapezoidal voltage is very well visible in Figure 49.

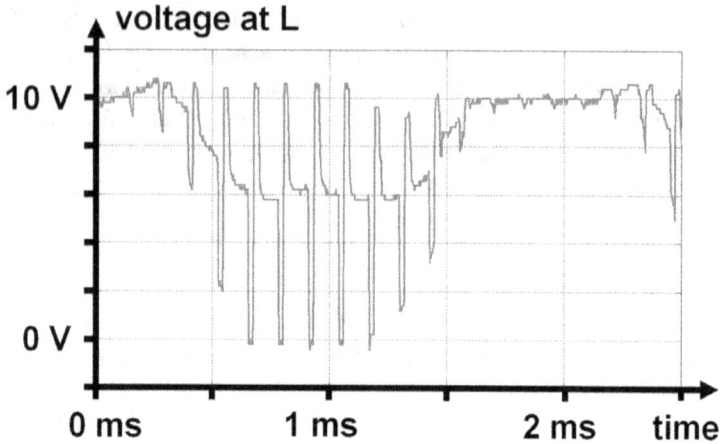

*Figure 49: Measurement of the voltage curve at one conductor*

The motor speed can also be read from the diagram. A period lasts about 2.1ms, which gives a frequency of 1 / 0.0021 Hz = 476 Hz. The outrunner has 14 poles, so the rotation frequency is smaller by 14/2 (see Chapter 2.4). The motor therefore rotates with 476/7 Hz = 68 Hz, and a rotational speed of 68 · 60 rpm = 4081 rpm.

The speed can also be read out in a different way from Figure 49. The average voltage is about 6V between 0.5ms and 1.5ms. The difference to the fully switched through upper transistor is 10V - 6V = 4V. The speed is equal to kV x $U_{BAT}$. kV is at the present motor 1060 rpm/V, and here these 4V represent $U_{BAT}$, i.e. n = 4 · 1060 rpm = 4240 rpm. This number doesn't match completely with the first one, which was calculated via the rotation frequency. On the one hand, the values are only read off from the figure and on the other hand the value from the calculation via the kV must be larger anyway, since here no idle current is taken into account.

## 4.2 Timing and phase measurement

After turning on the model, the drive is always situated in standstill and must always be started up to the desired speed. This can be done in different ways.

**Revving up the motor without a feedback signal of the speed and rotor position**
There are now several ways in which the controller can do this. It might start with a trapezoidal voltage with a small average and low frequency and slowly increase both voltage and frequency. Another possibility is to firstly apply a strong voltage pulse to one or more conductors in order to bring the rotor to a desired position. In motors with coils in slots a different inductance is the result, depending on the position. Some of the start-up algorithms measure the profile of the current depending on the position and give different voltage pulses on the conductor. After this initial positioning, the rotor is then operated with a three-phase trapezoidal signal with a growing voltage and frequency. This raises the danger here that the controller wants to start up the motor too quickly. This can happen especially when the unknown load torque is too large because then the rotor is lagging behind, as described in Section 2.4. Then, the rotor can fall out of step. This means that it stands still further, while the controller believes that it is already up to speed. In practice, this revving up method is used only rarely because of this disadvantage.

**Revving up with feedback of speed and rotor position**
The keyword for this is 'sensor-guided brushless motor', 'sensor-guided synchronous motor' or 'sensor brushless technology'. Particularly in industrial brushless motors, additional sensors are often used to detect the rotor position. There are, depending on the application, Hall sensors, quadrature encoders or resolvers. That is the silver bullet of revving up a brushless motor. The controller knows at any time the rotor position and speed and can react immediately if any of the values do not match with the calculated. Accordingly, the variant is more expensive, because additional sensors are required.

In the following, the Hall sensors will be discussed. Figure 50 shows the operating principle.

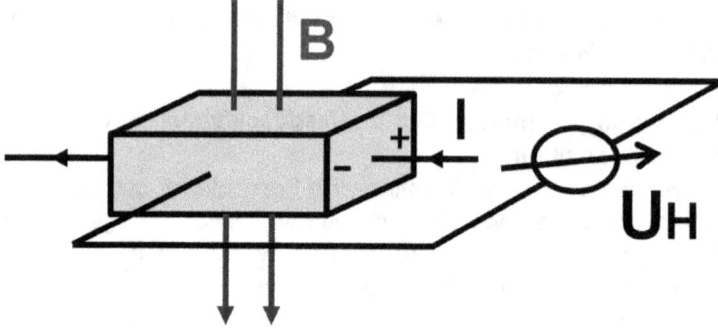

Figure 50: Hall effect

A semiconductor platelet will be flown through, by an electrical current. Perpendicular is a magnetic field. Unlike the discussion of the force acting on an electrical conductor (Fig. 10), the platelet is not pushed away here, since it is not movably mounted. Instead the negative charges collect on the left, seen from the technical current direction. This generates a Hall voltage that is proportional to the magnetic field.

Figure 51: Hall sensors in a brushless motor

Figure 51 shows how this physical principle is used for the detection of the rotor position of a two-pole inrunner.
The three plates in black housings are mounted on the stator 120 degrees to each other and measure the magnetic field of the rotor. Figure 52 shows a course of the amplified sensor signals for a turn.

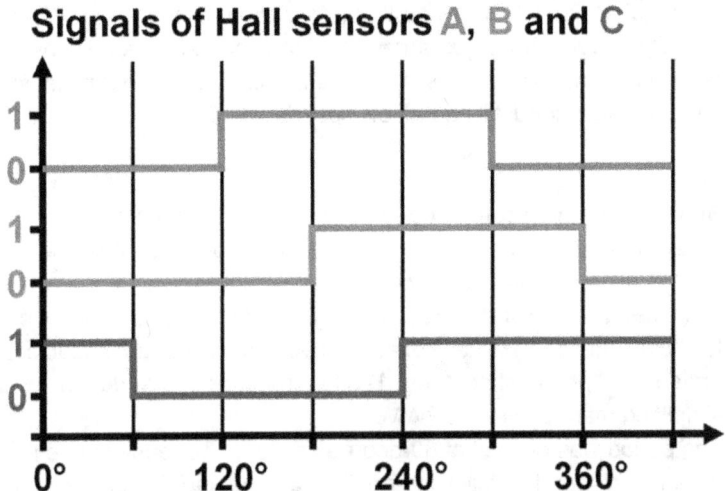

Figure 52: Logical signals of the three Hall sensors for a turn

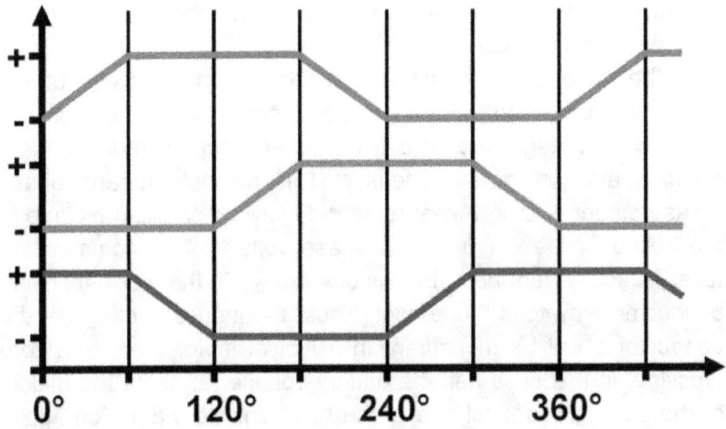

Figure 53: Trapezoidal voltage curves of all three conductors

They measure, depending on their position, a north or a south pole. With this two-pole rotor, each sensor has a resolution of half a turn, so 180 degrees. Through the use of three sensors, the resolution can be increased to 60 degrees. A higher number of poles, for example a four-pole rotor, increases the resolution further. That's exactly the resolution which is required to switch to the next phase. Special start-up algorithms can be omitted, since the regulator can always use this information to energize the right coil. The revving up is guaranteed under all operating conditions. In contrast to the next discussed variant, this measurement method also works at standstill and at low speeds.

**Sensorless revving up, sensorless brushless motors**
This variant is discussed in a bit more detail because it is used in almost all common model construction motors and many small industrial drives. For this purpose, Figure 53 should be considered. It shows the trapezoidal voltage curves of all three conductors. These are phase-shifted by 120 degrees and are drawn in a simplified manner without PWM.

The period was divided at 60-degree sections. It is easy to see that for each section the following applies: a conductor is connected to the positive battery voltage, a conductor is connected to the negative battery voltage and a conductor is not connected either to the positive or to the negative battery voltage, its voltage is increasing or decreasing.

It is the latter of these which has helped the success of the trapezoid voltage form. If a conductor is currently not connected to any battery voltage, then it can be traced back for measurement reasons and serve as a sensor. This principle means phase measurement or $U_{EMF}$ measurement. Figure 54 shows the curve of the voltage on a conductor, a phase voltage. The commutation takes place when both transistors block, if the corresponding conductor is switched to sensor. Thus, the induced voltage at the conductor L will be measured. In a typical motor there is a zero crossing, in this case half the battery voltage, at about the middle of the 60 degrees, until one of the two transistors turns on again, i.e. at 30 degrees.

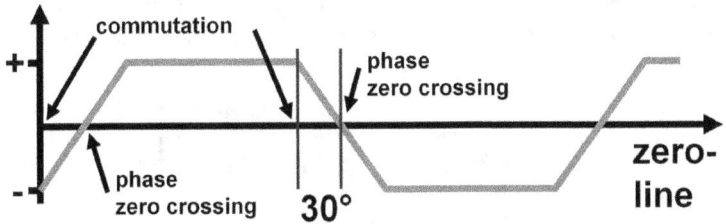

*Figure 54: Phase voltage with commutation and zero crossing*

The zero crossing of the curve is nothing other than information about the current position of the rotor. It tells the brushless controller when to switch to the next phase, i.e. when the next commutation should take place. If the zero crossing is before 30 degrees, then the controller must slightly enlarge the frequency of the trapezoidal voltage. If the zero crossing is after 30 degrees, then the frequency of the trapezoidal voltage is too high and it needs to be slightly reduced.

Actually, this principle is not 'sensorless', as there is measurement and feedback to the controller. But in technical language, this notion has prevailed everywhere where there are indeed measurements of the state of a system, but where no specially appointed sensors are required for it. This is also true in this case. It is the coil itself which for a short time works as a sensor and which for the remaining time is responsible for driving the motor. 'Sensorless' of course always means that it is an inexpensive solution in terms of price. There are indeed ultimately parts used which perform the dual function of measuring and driving.

The phase zero crossing of Figure 54 is measured with the circuit in Figure 55. The zero line is nothing more than the virtual star point. This must be reproduced in the controller with a resistor network, as only the three connectors L1, L2 and L3 will be traced back from the motor. Furthermore, all the measurement points have capacitors to ground. Since because of the switching frequency the measurement results are usually not as nice as in Figure 54, they assume the shapes as in Figure 49. They must therefore be filtered by a series circuit of resistor and capacitor. All signals, the virtual star point and the filtered measurement results zero1, zero2 and zero3 are led to the microprocessor. This knows the actual commutation state, can therefore compare the proper voltage curve with the virtual star point and thus determine the phase zero crossing.

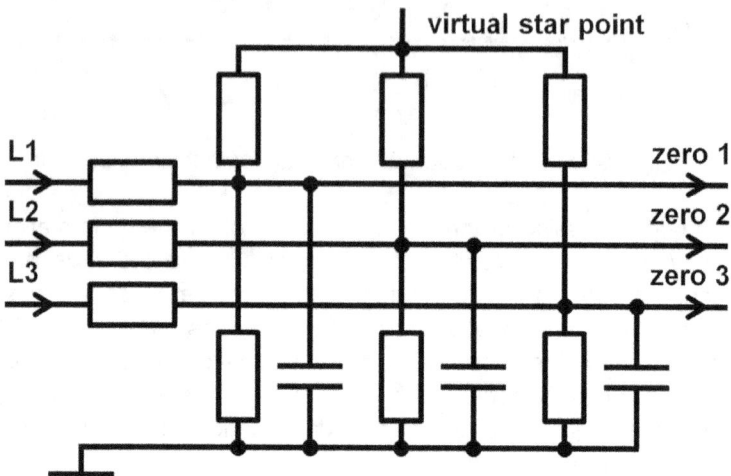

*Figure 55: Electric circuit for phase zero crossing measurement*

In almost all modes of operation, the brushless motor can be driven sensorlessly. But at standstill, this doesn't work. As mentioned in Section 3.1, and in particular in the statements of Figure 27, no voltage is induced at a standstill, when the magnetic field exerts no relative motion to the conductor. Thus, the controller must rev up the drive at first without the rotor position feedback signal, until a particular relatively low speed is reached. Then the $U_{EMF}$ measurement can take over.

**Timing**
But in reality the phase zero crossing occurs not exactly at 30°, but a little later. It was already discussed in Section 2.4 that the rotor under load torque lags slightly behind the current curve. The current curve lags behind even the trapezoidal voltage because of the motor inductance. So there are two factors responsible for the lagging behind of the rotor and thus the phase zero crossing: the motor inductance and the load torque.

Countering this 'lagging behind' of the rotor has introduced the term 'timing' into practice. If the phase zero crossing occurs as an example only 40° after the commutation, then the controller can compensate for that by undertaking the commutation at 40° – 30° = 10° earlier. The next phase takes over in this case by 10° earlier or, in other words, the controller has a timing of 10°.

The timing is in all of today's controllers one of the most important parameters. It can usually be set to a constant value. Some controllers can also automatically determine it (see example 1). For an inrunner with a small motor inductance and small load a timing of 5° is often chosen. For higher loads it can be up to 15°, because then the rotor is lagging even further behind. Outrunners with a larger motor inductance are usually driven with timings of about 15°, and in extreme cases and at high loads up to 30°. Since the motor inductance increases with the number of poles, one often hears in practice the rule of thumb:

**Timing in ° = 2 x pole number of the motor**

However, this doesn't take into account the motor load. A larger timing also means that the drive accelerates and decelerates even more strongly. Therefore, it is in practice also called 'hard timing'. A soft timing means the opposite: a smaller timing of a few degrees and a soft brake and acceleration. A motor which is operated with hard timing runs even with a smaller efficiency and requires a higher current. Anyone who wants to achieve higher efficiencies rather selects a soft timing and takes into account that the motor doesn't accelerate as strongly.

In many motors the recommended timing is described in the datasheet. It is therefore often given a range. If a small load is expected, so a low torque, rather the smaller (softer) timing should be selected. With expected high motor loads the larger (hard) timing should be set.

**1$^{st}$ example:**
*With a conductor of a brushless motor with a kV of 1060 rpm/V the voltage curve arises as shown in Figure 56. Based on the figure,*

the speed of the motor, the pole number and the timing should be determined.

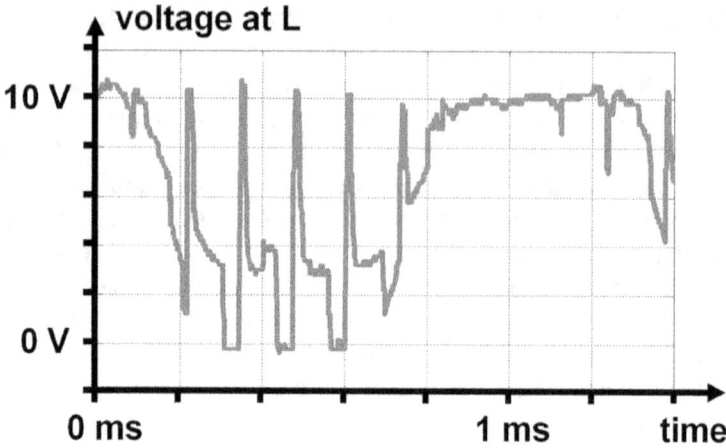

*Figure 56: Measured voltage curve at a conductor*

The difference between the upper and lower voltage average value is about 10V – 3V = 7V. The speed is therefore calculated as 7 x 1060 rpm = 7420 rpm. This corresponds to a rotation frequency of 7420 / 60 Hz = 124 Hz, a period is read out from the chart to about 1.2ms. This corresponds to a frequency of the alternating current of 1/0.0012 Hz = 833 Hz. 833 Hz is about seven times larger than 124 Hz, thus the number of poles is 2 x 7 = 14. The timing calculated from the rule of thumb is 2 x 14° = 28°.

The motor is a Hacker A30-16-S, with a recommended timing of 20° to 25° on the data sheet. The calculated 28° is thus a rather hard timing.

**Trapezoidal voltage and frequency during revving up**
After the $U_{EMF}$ measurement or the measurement with the Hall sensors come into action, the rotor position and speed are always known for the controller. If it sets a particular trapezoidal voltage, or more precisely a certain voltage difference between the average of the upper and lower voltage (in 1st example, the 10V - 3V = 7V), the motor is initially not in equilibrium and it rotates more slowly than the controller would like. This is shown in such a way that

after the equivalent circuit of Figure 29, the conductor voltage $U_L$ is larger than the induced voltage $U_{EMF}$. The $U_{EMF}$ is precisely proportional to the speed, which is still slow. Thus, there is a voltage difference, which is present across the winding resistance $R_W$, and therefore a large current flows through the coil. The current is always coupled with the torque. Since this torque is greater than what the load demands, it uses this torque for acceleration to the difference voltage corresponding speed. This corresponds about to the kV (calculated with no load) x the difference voltage.

Figure 54 of the timing curve is true only in the form of a constant speed. During the acceleration phase, the phase zero crossing is earlier due to the larger torque. Together with the timing, the controller automatically moves the frequency up to the frequency corresponding to the trapezoidal voltage. The higher the speed, the larger the induced voltage $U_{EMF}$. Thus, at a constant trapezoidal voltage, the voltage drop across the winding resistance will get smaller. Thus, the current and also the torque used for the acceleration is always getting smaller. At the end arises again with the final rotational speed an equilibrium between the trapezoidal voltage and the induced voltage $U_{EMF}$.

In practice, the controller is not revving up the trapezoidal voltage suddenly to the maximum voltage, but only after a delay. In many controllers, this delay can also be set. Together with the hard or soft timing, it determines acceleration and braking values.

## 4.3 Maximum current, current and temperature measurement

The maximum current is one of the most important parameters of a brushless controller. It is mostly used as a suffix. As an example, consider a controller of Kontronik. The 'KOBY 70 LV' can deliver a continuous current of 70A. This means the current which flows from the battery to the controller. On its way to the motor it always first flows through the MOSFETs. When these are switched through, they have a low resistance, the 'RDS_ON', which can be found in many data sheets of transistors. 'ON' stands for 'switched through', and 'RDS' for 'the resistance between drain and source connectors'. This resistance value generates a voltage drop by the current flowing through it and hence a loss of power $P = R \cdot I^2$. The MOSFET should therefore be cooled by a good connection to the circuit board so that it can dissipate heat generated by the power. Many controllers have small PC fans; some even have a water supply with pump to better handle the heat problem.

As practical advice, a controller should not be driven in continuous operation with more than 70% of the specified current. It will repay this consideration with lower heat generation and a concomitant longer life span. The author would operate a KOBY 70 LV only with a continuous current of about 50A.

Whereas the output stage and the measurement of the phase zero crossing belong to the minimal necessary equipment of a brushless controller, the current measurement is not in any case part of it. This concerns additional information about the motor current. Thus, the controller can determine if it is already operating in the overload range. The controller can do many things with that information; some behaviors due to the current can also be set with the parameters. In the simplest case, it gives out a warning sound at too high a current. There is also the possibility that it drives back the speed a little or even completely shuts down the motor. But this is not desirable for all applications. As information about the current always means information about the torque of the motor, some controllers automatically adjust the timing.

*Figure 57: Shunt of a brushless controller*

Figure 57 shows the controller of a brushless quadrocopter and Figure 58 is a schematic illustration.

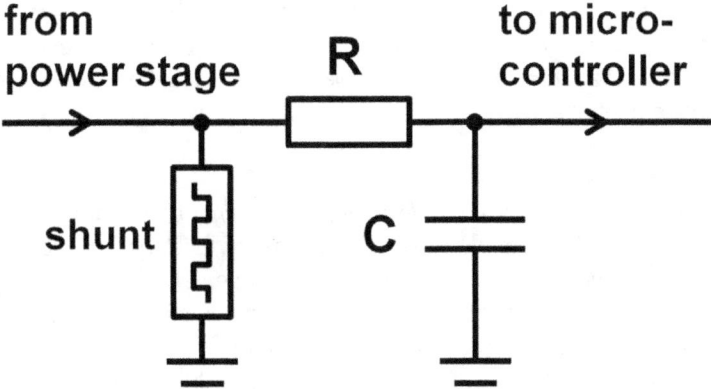

*Figure 58: Electrical circuit of a shunt*

The wound conductor has together with the thickness of etched copper the function of achieving a certain defined small resistance. The current flowing through it then generates a voltage drop according to the law U = R · I. The resistance of the shunt is typically approximately in the range of 0.01Ω or even smaller. Consider this: If a 50A current flows, then at a shunt of 0.005Ω there is already a voltage drop of 50 · 0.005V = 0.25V. This voltage drop reduces the applied voltage of the motor. Before the output of the shunt is led to the microcontroller for measuring purposes, it is often still filtered by a resistor and a capacitor.

**Temperature measurement**
Power dissipation always results in heat development. It is in the end what causes the overload of the controller, because the electronic components are then operated above the permitted temperature. Often it is measured in addition to the current, but sometimes instead of this. Frequently, the temperature dependence of the resistance of platinum is used. The sensors are then called, for example, PT100 or PT1000. They must then be placed where the greatest heat is generated, i.e. close to the MOSFET. For controllers which are equipped with a temperature

measurement, often parameters similar to the measurement of current can be set, such as a shutdown or reduction in rotational speed at a high temperature.

The current and temperature measurements are redundant. At high currents the temperature will also be higher. Vice versa, at high temperatures the current is also high, unless there is a defect in the controller.

**$2^{nd}$ example:**
*A brushless controller would provide a continuous current of 30A. The shunt for current measurement has a resistance of 0.005Ω. The used MOSFETs each have a resistance value of $RDS\_ON$ = 0.005 Ω. There the losses in the power stage should be calculated. The current flows on its way first through the upper MOSFET, then through the motor winding and then through the lower MOSFET, before it flows through the shunt. The loss in the power stage is then $(0.005+0.005+0.005) \, \Omega \cdot 30A \cdot 30A = 13.5W$.*

It was already mentioned in Section 3.3, the rotational speed at large motor currents drops slightly more than the mere internal resistance of the motor would allow. The resistor values of the controller mentioned in the example must be added in practice. Although they are usually smaller than the winding resistor, they can nevertheless not be neglected. The resistance of supply cable and the battery internal resistance must also be added to get the true resistance.

Completely neglected in this calculation were the switching losses, which were discussed earlier. They also cause additional loss in the power stage.

### 4.4 BEC voltage

As the measurement of current, the BEC voltage is also something that needs not necessarily be a part of a brushless controller. But it is usually installed in model construction applications. The term BEC stands for "Battery Eliminator Circuit". Of course the system doesn't work without any battery. It is thus rather meant that it is not necessary to have two different batteries for the motor and for the receiver with the servos. It would have the advantage of redundancy if the receiver and the motor were supplied separately. Namely in almost all the models, the function of the receiver is much more important than that of the motor. It must therefore be a higher priority, which means that the receiver still has to work when the motor was shut down due to low battery voltage. But especially with the flight models installing two batteries in the model adds additional weight. Instead, the BEC is used. There are two implementation types, each with advantages and disadvantages.

**BEC with linear voltage regulator**
Many more simple BECs are equipped with so-called linear voltage regulators. A linear voltage regulator is an electronic component which produces a constant voltage of 5V or 6V when the battery voltage is larger than about 7V. If the receiver is now pulling a current with its servos, then the linear regulator dissipates the surplus power, which results from surplus voltage x current.

**3$^{rd}$ example:**
*A model airplane will be operated once with a 2S LiPo with 7.4 volts and once with a 3S LiPo with 11.1 volts. The BEC is designed with a linear voltage regulator and must provide 5 volts for the receiver. This pulls together with the servos a current of 2A. The resulting power dissipation in the BEC should be calculated for both cases.*
The following applies for the first case: $Pv = (7.4V - 5V) \cdot 2A = 4.8W$. For the second case $Pv = (11.1V - 5V) \cdot 2A = 12.2W$.

As the example shows, the power loss increases with a larger number of cells, since there is then a larger surplus of the voltage compared to the 5V or 6V of the receiver. This can especially be a problem if the receiver and particularly the servos pull a very high current and in addition the number of cells is large. Even with a power loss in the range of 10W to 20W, the power loss of a BEC is a significant percentage of the controller losses and thus its heat development. With car models in which the steering servo often pulls hard power, caution is necessary with such BECs.

BECs performed with linear voltage regulators also have an advantage in comparison to the switching voltage regulators treated later. Their voltage provided for the receiver is of very low noise.

### BEC with switching voltage regulator, SBEC

Thanks to today's technology, model construction has the tendency towards ever-larger models with higher power. The BECs with linear voltage regulators are only recommended for smaller models and a small number of accumulator cells. There, however, they quite have their purpose.

Large models often have a very large number of accumulator cells. Their servos also pull a high current. Thus, a BEC equipped with a linear voltage regulator would generate far too much power dissipation and heat. Therefore BECs with switching voltage regulators are used here. The switching voltage regulators operate – simply explained – somewhat like transformers, but with the difference that not AC voltages and AC currents but DC voltages and DC currents are transformed. They work (nearly) losslessly. The power provided at the output of the controller is (almost) the same size as the power at the input.

### $4^{th}$ example:

*A large model would be operated with a 10S LiPo at 37V. The receiver needs a voltage of 5V and pulls a current of 2.5A together with the servos. The BEC is designed with a switching voltage regulator. It should be calculated which current the power supply requires from the battery.*

*The power of the receiver electronics with servos is 5V · 2.5A = 12.5W. Since the power at the input of the BEC is about the same as at the output, the current is 12.5 W / 37V = 0.34A.*

*Thus, the nearly losslessly switching BEC generates almost no heat. By comparison, a BEC with linear voltage regulator would take 2.5A from the battery and would generate the unacceptable power loss of (37 – 5) V · 2.5A = 80W with corresponding heat.*

As the name suggests, the switching BEC, often also called SBEC, is made with switching transistors, and chops the input voltage.

Therefore it requires a slightly higher effort to provide a low-noise and interference-free receiver voltage.

**Control voltage**

The BEC always has three connectors. These are the standard servo connectors Ground, 5V and the control voltage for the regulator. It provides on the one hand the receiver with the Ground and 5V, as mentioned above. With the control voltage, the brushless controller gets in return the information about the position of the transmitter stick. So it is also this plug which gives it the command to set a higher or lower speed.

This signal is also a pulse width modulated (PWM) signal. It has been standard for years in model construction. Figure 59 shows the course of the control voltage.

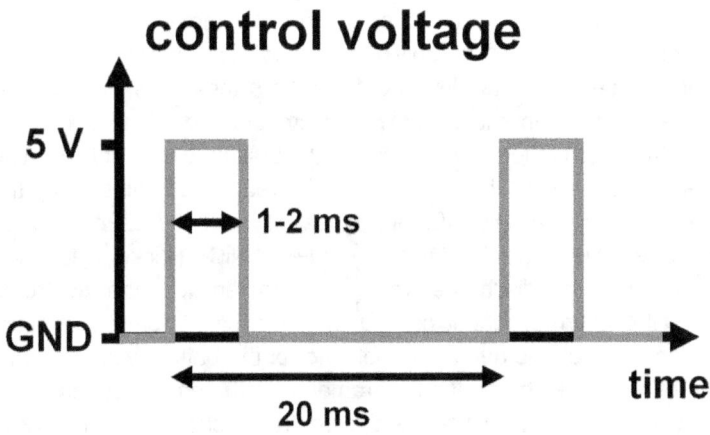

*Figure 59: Control voltage, PWM*

The signal is repeated every 20ms. At the very beginning, it changes to logical high, so 5V. There it remains for 1ms to 2ms, depending on the stick position. 1ms means 'low speed' or 'standstill'; 2ms means 'maximum speed'. Of course it can also be reversed, depending on whether the signal is inverted at the remote control or not.

The pulse width modulation with the repetition of 20ms is nowadays no longer fast enough for all possible applications. With the modern quadrocopters for example, a fast reaction speed is important, so that they can keep stable in the balance. Here there is on the one hand a variant of a faster PWM with a faster repetition of the signal. But on the other hand another type of communication is often used, a so-called I2C bus, pronounced 'I-squared-C-bus'.

## 4.5 Low voltage measurement

After a brushless drive brings an electric glider to a height, then it must be guaranteed that the receiver can allow thermic gliding for an hour or even more with the low battery. Even with a car model with a LiPo battery, it is important that the motor is switched off before the battery is deeply discharged. This particular type of battery doesn't like too low a voltage. With these examples proper battery management is of prime importance. Again, it's the brushless controller which performs the task of 'power station'. The battery voltage will therefore always be read and evaluated by the controller. Since the different types of batteries LiPo (Lithium Polymer), NiMH (Nickel Metal Hydride) or NiCd (Nickel Cadmium) show different discharge voltage characteristics, the controller must be informed about cell number and type. Some controllers recognize the type due to the behavior of the battery voltage, while for others it can be set as a parameter. With some systems it can also be set at which voltage the controller should switch off the drive or issue a warning via a flashing light or buzzer.

In any case, one should study the manual carefully when it concerns battery management. If one thing is set wrong here, the simplest case of a fault is a possible battery deep discharge. If the receiver loses contact to the transmitter due to incorrectly set parameters, sometimes even the loss of a model may be the result.

## 4.6 Operation modes, stick position = voltage or speed

At the time when controllers were still driving brushed DC motors, the controller gave out a pulse-width modulated DC voltage signal whose average value was proportional to the idle speed of the drive. The average value of the voltage signal and thus the idle speed was therefore proportional to the stick position of the remote control. This is also the same with the use of brushless motors. The only difference is that only the trapezoidal voltages are issued on the conductors. As the motor characteristics of Section 3.3 show, the speed drops slightly under load. This may not be a problem for many models, because it ultimately belongs to the nature of each motor. Therefore, this mode is the default configuration of the brushless motors.

**Helicopter mode, governor mode**
The information about the speed is available by measuring the phase zero crossing, and therefore with a brushless motor and the appropriate controller even another mode of operation can be used, the so-called governor or helicopter mode.
If the speed is known, one can also imagine that the controller is configured so that not just the trapezoidal voltage but also the actual speed itself is kept constant independent of the load torque. This is specifically interesting for a pitch-controlled model helicopter. The controllers are here equipped with various software solutions for the realization. In one configuration possibility the stick position corresponds to the actual speed. In another, a specific speed can be preset. After passing through a given stick position, for example the center position, this speed is strictly adhered to, regardless of the pitch position of the rotor blades.

## 4.7 Microcontroller and setting the parameters

The previous subsections have shown that brushless controllers have to cope with far more responsibilities than just controlling the motor. They are also responsible for the proper energy management, for battery warnings at low voltages or for monitoring the motor current. All this makes it necessary to be able to set a variety of parameters. There are different possibilities how this can be done. The possibilities are as follows, each of which will afterwards be discussed in a separate sub-chapter.

- Programming via the remote control, buttons, jumper or buzzer
- Programming via a programming box
- Programming via a USB interface and PC software

In order that a controller can be programmed at all, a microcontroller will be required. Already in Figure 45, the block diagram of the controller, it is evident that all the important information and control lines are connected to it. Without exception, all brushless controllers include a more or less fast microcontroller which handles all tasks and which mostly can also be configured or programmed by the user.

**Programming via the remote control, buttons, jumper or buzzer**
This programming method is the easiest possibility. It is not only used for small controllers with few setting options, but also for controllers with a high degree of automation. There the software detects the different modes of operation automatically due to the motor behavior and sets the parameters almost without user interaction.

Some systems don't have any buttons. Here the programming mode is achieved via a particular position of the remote control stick, when the brushless controller is connected to the receiver. The controller then acknowledges the parameter settings with a sequence of beeps, as specified in the operating manual. With the stick of the remote control the parameter values then can be set.

If the manufacturer has granted the controller a button, the parameter values can be entered by a sequence of button-presses. As an example, if after the first long beep the strength of the EMF-brake is asked, it can be acknowledged with one, two or three button presses, depending on the desired setting.

In several systems, parameter settings can also be made in an easy way with a jumper.

**Programming via a programming box**
With controllers with more options for the parameterization, a programming box as in Figure 60 is often used. Other names are also 'programming card' or 'programming interface'.

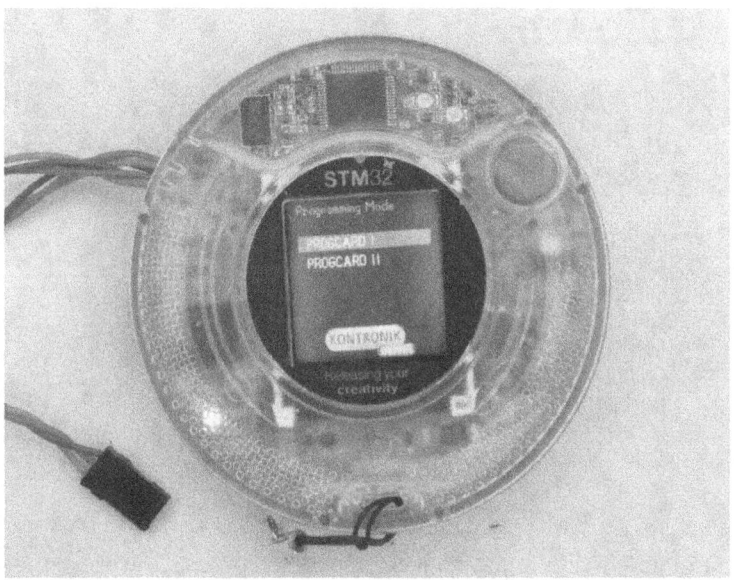

*Figure 60: Programming box*

As can be seen, the card is often connected to the BEC plug. It now doesn't send a control signal as shown in Figure 59, but a sequence by the manufacturer defined of 5V and 0V. The controller detects this and switches to the programming mode. This is of course much more convenient than in the variant with buttons and remote control. Each of the submenus, the low voltage shutdown, for example, can be selected and things such as cell number or type of battery can be entered. Often, the manufacturers also provide a combination of buttons and programming box. Thus, with the programming box all possibilities are covered, whereas the buttons support only the most important settings.

**Programming via a USB interface and PC software**

In many controllers, the programming is realized via a USB interface and the PC. The connecting cable can often be plugged directly into the controller. In model construction it is also usually the BEC through which the connection is made. Therefore, many manufacturers also provide a small converter print from BEC to USB and deliver their controller with software on CD. Alternatively, this can also be downloaded from the Internet. Figure 61 shows such a converter print.

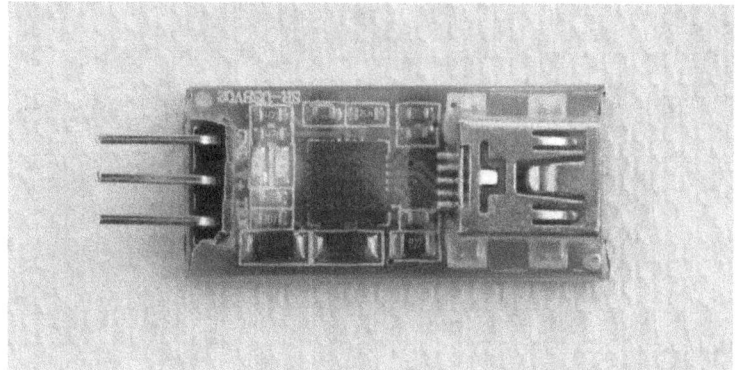

*Figure 61: BEC–USB converter print*

On the left side is the connector plug for the BEC, while on the right side the USB interface can be connected from the PC. With a programming interface on the PC, now the same functions as with the programming box can be realized. It is a kind of programming box on the screen. The converter print is cheaper to produce than the programming box. This might be seen as an advantage. A disadvantage, however, is that you must always have a computer at hand to make changes in the parameterization. In model construction the parameterization takes mostly place on the airfield, on the shore of a lake or next to the race track. There, a programming box is somewhat easier to handle than a computer. But both solutions are equivalent, and many manufacturers are offering both options for their controllers.

**Non- volatile memory**
It is standard for all controllers that the parameters, once set, remain stored even after the decoupling from the battery. All of today's storage devices are so-called non-volatile. The same technology is used here as in the computer world with the popular USB memory sticks. In practice, however, one sometimes hears that parameters have been lost. Who wants to feel confident, should briefly test them before the start, especially if he has previously not operated the drive for a while.

**Shutdown analysis**
There are many possible reasons for the controller to reduce the speed of the motor or even to shut it down. The battery voltage may be too low, the motor current may be too high, the controller may be too hot, or in an extreme case, even the measurement of the phase zero crossing may fail due to a defective contact or a bad configuration. Here it is very desirable that the user can learn about the reasons afterwards. There is no feeling less secure than not knowing about the causes of a failure and the certainty that this can be repeated any time. Various controllers provide a shutdown analysis for this. This is the same as a black box in a large aircraft, but made much simpler.

In the most simple case, there are LEDs on the controller which represent various error cases in the operation by the pattern of flashing. You can then identify the error using a table in the manual. However, there is with more complex controllers also the possibility of connecting a PC. Then even the time courses of the signals can be followed up just before and during the shutdown. This may be of great importance in the analysis of the fault. Depending on the course of the signals, corresponding measures may be taken. It helps to avoid in the future the errors that occurred.

## 4.8   Findings in brief

In the following, the main findings from Chapter 4 are summarized.

*The brushless motor is driven with a trapezoidal voltage. Here, during 120 degrees the upper transistor is switched through and during 120 degrees the lower transistor is switched through. During the remaining 2 x 60 degrees of a period, both transistors block and the conductor can be used for the measurement of the phase zero crossing. For part load operation and for revving up the motor, smaller voltages than the battery voltage can be realized with pulse width modulation.*

*The switching frequency is usually between 8 kHz and 32 kHz. Low switching frequencies are selected in motors with large motor inductance (typically outrunners). High switching frequencies are selected for motors with smaller motor inductance (inrunners with iron-free coil).*

*The timing angle refers to the commutation which took place earlier.*
*Hard timing means: large angle, high current, faster revving up.*
*Soft timing means: small angle, small current, slow revving up.*
*The motors used in model construction usually work in the sensorless mode. Here, the rotor position is recognized when both transistors of the corresponding phase are turned off.*
*At standstill and at low speeds, the sensorless operation doesn't work because of the lack of the $U_{EMF}$ measurement. The motor must be driven to a low speed with a special start-up strategy until the measurement intervenes.*

*There are also motors whose revving up is driven by sensors. In this case, Hall sensors are often used as feedback. This allows a jerk-free start, because the rotor position is always known, even at low speeds.*

*Most brushless controllers are equipped with a current measurement. These can turn off the motor at too high a current. In practice, the controller should be somewhat oversized. In continuous mode, it should not need to deliver more than 70% of the current specified in the data sheet.*

*The BEC (Battery Eliminator Circuit) is used to power the receiver and servos by the motor battery. For smaller models and a small number of cells it is performed with linear voltage regulators. Larger models with a larger number of cells are operated with switched BEC, also called SBEC.*

*The low voltage shutdown should always be studied accurately in the datasheet of the controller. If it is set incorrectly, a total discharge of a battery or even the loss of a model by a receiver error could be the result.*

*For simple systems with few setting parameters, the configuration or programming of the controller is made via the transmitter or a button. For controllers with many setting parameters, a programming box or a BEC–USB converter print with PC software is used. In some systems, the USB cable of the PC can be plugged directly into the controller.*

*The memory of today's brushless controllers is non-volatile. This means that all parameters remain stored after disconnecting the battery.*

## 5. Practical examples

In this chapter, the above explanations will be applied to examples from model construction. We begin by giving some practical advice for cabling. Then a motor and propeller combination is measured with a dynamometer. The modification of an RC monster truck model from brushed to brushless follows.

### 5.1 Practical tips for cabling

**Long cables**
The interference of brush motors was reduced with capacitors and inductances, because the mechanical commutation was resulting in short current interruptions. These led to voltage spikes and the so-called brush fire. This also caused disturbances at all frequencies. The brushless motor is almost clean in this regard. As long as the battery, motor, controller and cable length specified by the manufacturer are used, no further steps need to be taken to reduce interference.

But often it is necessary to use longer cables. There are two possibilities, namely an extension between the battery and controller or between the controller and motor. The cross-section is selected the same as the cables to be connected. An extension cable between battery and controller has the advantage that this happens on the DC voltage side. Therefore no interference occurs on the receiver. Since this cable is supplying along with the power stage also the BEC and the microprocessor itself, only pure DC should arrive. But with an extension, the cable itself acts as an antenna. Therefore, electrolytic capacitors (Elkos) can be found at the input of each controller.

In the case of an extension between the battery and controller, several capacitors with a small series resistance, so-called LOW ESR Elkos, should therefore be inserted if possible, as shown in Figure 62. Since Elkos can explode with reverse polarity, measures must be taken against this.

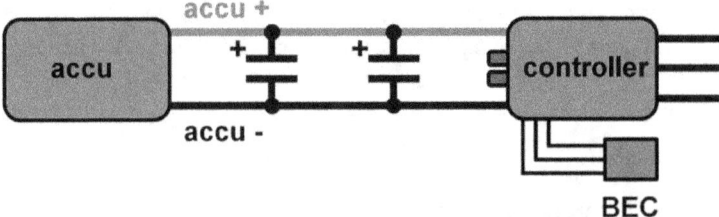

*Figure 62: LOW ESR Elkos in the cable*

The extension cables between the controller and motor are not critical with regard to their own antenna effect. However, because these cables have pulse-width modulated trapezoidal voltages, they can themselves cause harmful interference to the receiver. It is therefore also advisable to keep the cables as short as possible and to install the receiver and antenna as far away as possible. Capacitors between these cables must not be used, because they would endanger the correct interaction between the controller and motor.

**Ferrite core at the BEC cable**
In many controllers, the BEC cable is already equipped as standard with a ferrite core. Figure 63 shows this.

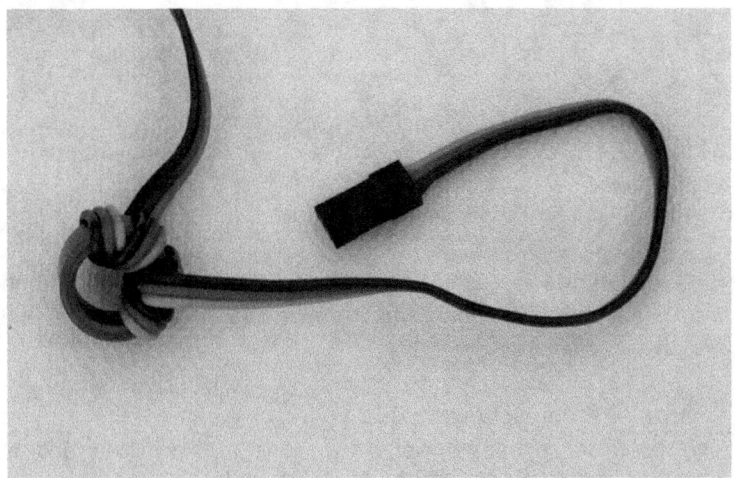

*Figure 63: Ferrite core at the BEC cable*

It dampens interferences, which come from the controller side, in order that the receiver gets a high quality voltage. At reception problems, this is also a possibility for better filtering. The cable is wrapped several times in the same direction around the core.

## 5.2  Graupner Compact drive

The first tested drive is a Compact 260Z at 11.1V from Graupner. This is a representative of the power class up to 100W. Its use is for example for slow flyers or other small model airplanes. The recommended controller for the Compact 260Z is the Compact Fly 15 BEC. For the drive combination, there is also a proposal of the propeller, namely a 7.5" x 4".

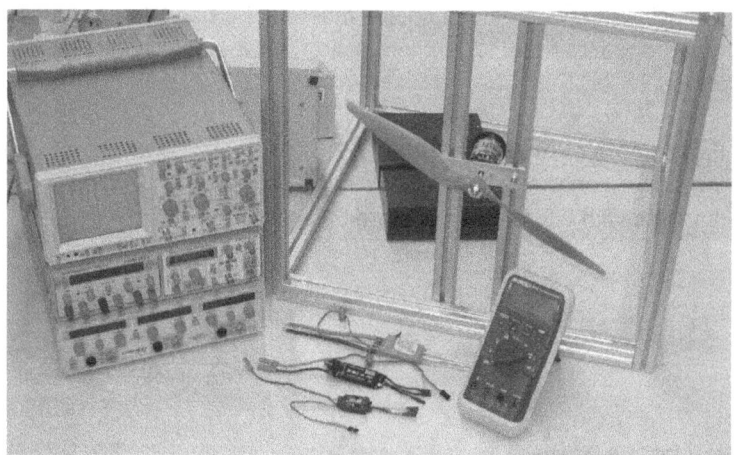

*Figure 64: Test bench and measuring equipment*

Figure 64 shows a test bench consisting of aluminum profiles and the measuring equipment, which can measure the voltage and current waveforms.

**Controller, Compact Fly 15 BEC**
The controller doesn't have many setting possibilities and is therefore very easy to operate. Since it is recommended for the

motors of the Compact series, it is perfectly adapted to their operating conditions. The number 15 stands for the maximum current and the BEC is the supply for the receiver, as explained in Section 4.4. The switching frequency is constant at 12 kHz, and the timing is automatically adjusted by the controller.

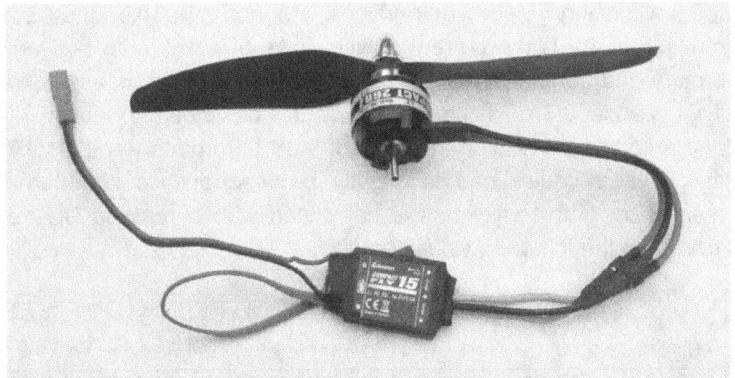

*Figure 65: Graupner Compact drive*

You can choose whether the motor brake should be on or off. According to the manual, this is accomplished with the throttle setting at the time of switching on the programming mode. The controller confirms its programming mode with beeps. These are emitted through the windings of the motor. If windings are energized, the individual windings push against each other and emit a tone at a level twice the frequency of the current. This principle can also be observed with transformers. The 50Hz of AC current buzzes at 100Hz. It is therefore twice the frequency, because the windings repel each other equally at positive and negative current. Although the motor is not an ideal loudspeaker, in this way even simple melodies can be played on it.

**Motor, Compact 260Z**
According to the data sheet, the motor has a maximum efficiency of 77%. Using the formula of Section 3.7, this is calculated to

$$\eta = \frac{P_{MECH}}{P_{EL}} = \frac{(U_{BAT} - I \cdot R_i) \cdot (I - I_0)}{U_{BAT} \cdot I}$$

So the internal resistance Ri and the idle current I₀ are required for the calculation. Since their values are not available, they are measured. As $I_0$ the current, which flows at full throttle between battery and controller without any mounted propeller, is assumed. That is about 0.7A. Ri is measured as it is described in Section 3.2 and amounts 0.212Ω. In the case of a 3S LiPo battery with 11.1V, the voltage under load is slightly reduced due to its internal resistance. If the efficiency is calculated with a realistic $U_{BAT}$ of 10V, then the following values result:

| Motor current | 4A | 5A | 6A | 7A | 8A | 9A |
| --- | --- | --- | --- | --- | --- | --- |
| Efficiency | 75.5% | 76.8% | 77.1% | 76.6% | 75.7% | 74.6% |

The value for the efficiency given in the datasheet will therefore be achieved and a reasonable range of operation are currents between 4A and 9A.

**Full throttle**
When the motor is operated with full throttle, the MOSFETs are not operated with pulse width modulation. They are always alternatingly switched through, as Figure 47 in Section 4.1 shows. Figure 66 shows the curve of the voltage between one of the three motor conductors and the minus pole of the battery for the above-described drive combination including propeller.
In comparison with Figure 47 again, the time periods with fully switched through upper and lower MOSFETs are identifiable. Even those areas where both transistors are open are very easily recognizable. There, the conductor is used to measure the phase zero crossing in accordance with Section 4.2.
Not mentioned in the theory were the pulses marked with a circle. They are found in every drive and always occur when the

MOSFETs are turned off. The motor inductance is responsible for it. It has the property that the current can't be stopped immediately, but still continues to flow a little. It must then temporarily flow through a diode (called a free-wheeling diode), which is in parallel to each other MOSFET. The conductor therefore has for a short time the other supply voltage ($U_{BAT}$ when the lower MOSFET is turned off or ground if the upper MOSFET is turned off). This voltage peak is also known in the technical literature as return pulse. The free-wheeling diodes are mostly already integrated in the MOSFET and don't need to be installed separately. It is less than the 3 x 3.7V = 11.1V at the conductor, when the upper MOSFET is switched through. Since in this example flows a current of about 9A at full load, a small voltage drops at both the internal resistance of the battery and RDS_ON of the power stage. Furthermore the battery pack is operated in the presented measurement rather in the second half of the term.

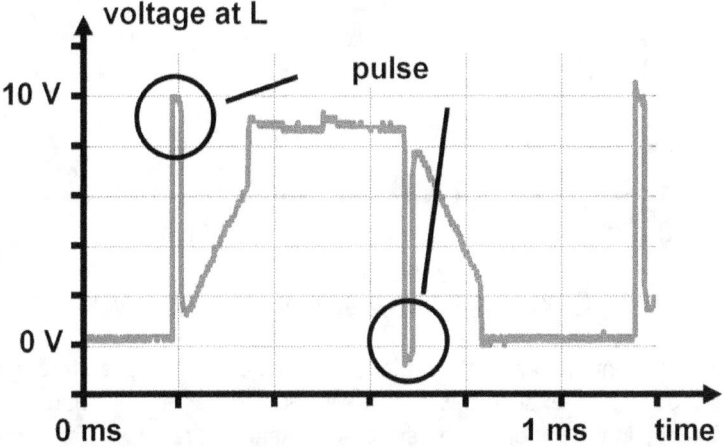

*Figure 66: Trapezoidal voltage waveform at one conductor*

The rotational speed can also be read from Figure 66. A period lasts approximately 0.97ms. This gives a frequency of 1 / 0.97ms = 1031 Hz. The Graupner Compact 260Z is a 14-pole motor. Thus it turns only with 1031 / 14 x 2 = 147 Hz. In the unit revolutions per minute it is still 60 times higher at 8837 rpm. Figure 67 shows the current through one of the three conductors under the same conditions. When the upper MOSFET is switched through, around 9A is flowing; if the lower MOSFET is switched through, -9A is flowing. The battery current is in this measurement constant at 9A. This is not a contradiction to the -9A of conductor current, since the power stage then reverses the battery current. During the measurement of phase zero crossing, the conductor is currentless. The round current curves shortly after turning on the MOSFETs are again due to the inductance of the motor. This is the visualization of the 'lagging behind' of the current, described in Section 4.2.

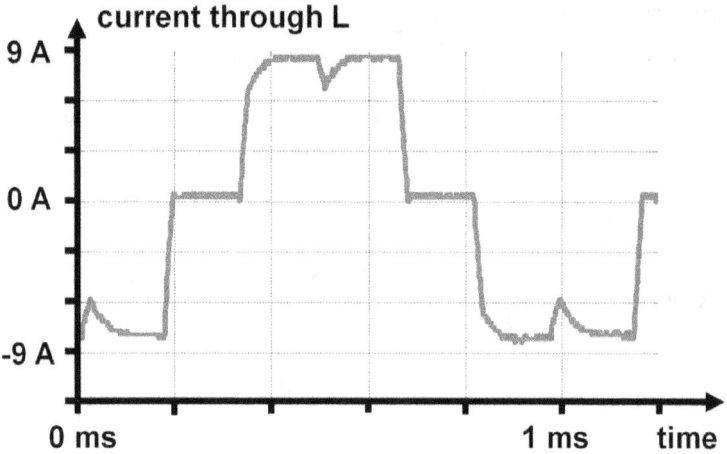

*Figure 67: Curve of current through a conductor under full throttle*

The current breaks a little bit in the middle of the switched through MOSFET. To explain this fact, we refer once again to Figure 53. Exactly in the middle of the switch-on phase of the upper transistor of conductor 1, the switch-on phase of the lower transistor changes from conductor 2 to conductor 3. Since the switching on and off takes some time, there is a short time during which neither of the two conductors 2 and 3 take up the current.

Both the voltage and current waveforms correspond with the theory, with the exceptions from the practice discussed above.

**Partial throttle**

The partial gas operation demands a little more from the controller, since the MOSFETs are driven by pulse width modulation. As a comparison serve Figures 48 and 49 from Section 4.1. Figures 68 and 69 represent the voltage and current waveforms at one of the three motor conductors.

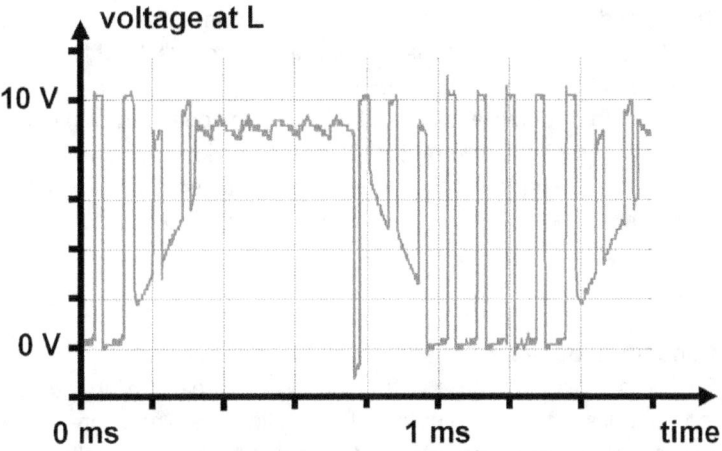

*Figure 68: Voltage curve at one conductor at partial throttle*

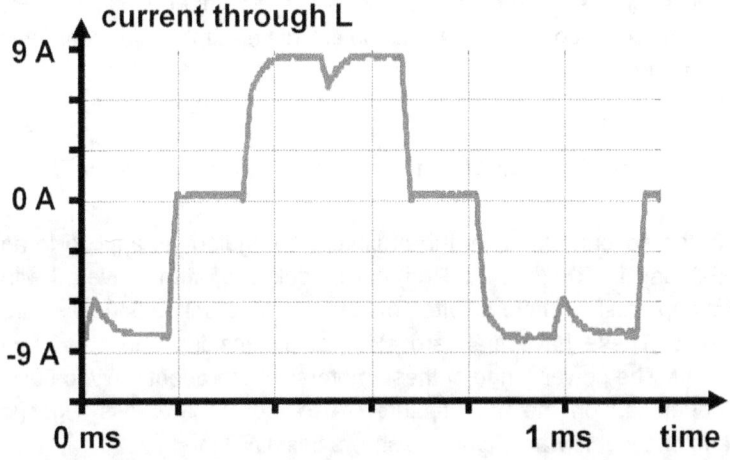

*Figure 69: Current curve through a conductor at partial throttle*

The conditions are the same as in the operation at full throttle, with the exception that the throttle is only about halfway from the maximum position.

The return pulse is also clearly visible. Moreover, the switching frequency of the controller may also be read off there. Within a millisecond it is switched twelve times, which leads to a frequency of 12 kHz.

The speed must logically be smaller than during operation under full throttle. A short calculation analogous to full throttle is 1 / 0.00125 / 2 x 14 x 60 rpm = 6857 rpm.

The waveform of current looks at every drive so jagged with partial gas. It is once again the motor inductance which causes this. The motor current and thus the current which is taken from the battery is the mean, which is the dashed line in the figure. It is for this example about 4.5A.

**Generation of heat**

For both operating cases, full throttle and partial throttle, the temperature after 10 minutes of operation was measured. The motor was thereby located in the air stream of the propeller. The controller was mounted protected from this, so just as if it had been packed in the fuselage of a model airplane without ventilation. For all operating conditions, the complete drive combination was only lukewarm.

### 5.3 Modification of a monster truck to a brushless motor

In the second example, the brushless technology is applied to an RC Car. RTR (Ready to Run) car models are often delivered with cheap brushed motors. Often motors of the so-called 540 class are used. These have standard external dimensions and a 3.2-mm shaft. The power range of these motors is from about 60W to 90W, depending on the manufacturer. Also the Reely Cross Tiger is motorized this way. Figure 70 shows this 4WD model.

*Figure 70: Reely Cross Tiger*

The manufacturers of brushless motors in the car model field offer spare motors of the same size for the 540 class. The difference in the power is shown here in an impressive way. Depending on the type, the brushless motor of the same size can be operated with up to 200W power. The higher efficiency is shown to be very effective here. The comparison of a brushed motor with 60% efficiency with a brushless motor with 80% efficiency indicates it. With the former, 40% of electric energy has to be dissipated as heat whereas with the latter only 20%. This leads to approximately double the power of the brushless motor of the same size.

**Dimensioning**
If you use a brushless motor, it must be dimensioned to fit on the mechanics. If you choose a type with a much too high a speed, the joy of the model will not last long, because soon the bearings will begin to wobble or the gear wheels lose their teeth. So one must first find out how large the speed of the original motor is. Unfortunately this is often difficult, because the information is usually not available. Measurements with a 7.2V NiCd battery pack give approximately 15,000 rpm. Figure 71 shows the original drive. Because of the low efficiency, it is provided with cooling fins.

*Figure 71: Original drive, a brushed motor*

LRP offers the motors of the type 'Eraser Sports Modified' with different turns. The version with 15.5 turns has a kV of 2800 rpm/V. Together with the 7.2V battery the idle speed is at 7.2 x 2800 rpm = 20,160 rpm. The power is given as 130 watts. As the pinion gear a gear wheel module 0.6, 19 teeth and 3.2mm bore size is used, as with the original motor. It is to be expected with these dimensions that the brushless motor gives the model on the one hand a better acceleration and on the other hand also a slightly higher top speed. The tests should show whether the mechanics of the Cross Tiger can withstand this.

**Sensor technology**
As discussed in Section 4.2, there are also in the model construction some applications with sensor-guided revving up of the brushless motor. The RC car models often have large variations in load torque. On the one hand, they must be able to accelerate from standstill with a maximum force. On the other hand, they have to brake down to a standstill with a maximum force. Thus, the use of this technology with sensors is offered here. Since both the motor and the controller must be designed for that, the combination of a sensor-equipped 'Eraser Sports Modified'

motor together with a 'Brushless Reverse Digital' brushless controller is used. Figure 72 shows the drive combination. Besides the three motor cables, the needed additional sensor cable can also be clearly seen.

*Figure 72: The built-in brushless motor with sensor feedback*

**Initial operation**
Also this drive combination is characterized by a high degree of automation. Thus it is delivered without any programming or interface card. In the manual, the user is just asked to firstly drive in the forward direction. Thus, the direction is detected, and the acceleration and braking torques are adjusted. Since the sensors detect the rotor position and the controller must energize the right coils with this information, the red, blue and yellow cables of motor and controller have to match together. The reversal of direction must therefore be made with the gas inversion of the remote control. That's about it; the rest is done automatically.

**Driving and measurement**

The Hall sensors do a great job in this application. Specifically, the startup behavior could never be controlled as well without them. The drive accelerates properly up to maximum speed. Even in the very low speed range, it can be controlled well. Even a slow movement with about 10 cm/second is possible. If the motor were sensorless, a phase zero crossing still couldn't be measured at this speed, because the $U_{EMF}$ voltage would be too small.

The brushless drive plays off its full strength in comparison to the brushed motor during the acceleration and deceleration phase and in the top speed. Figure 73 shows the measurement of the maximum speed on a paved road with a GPS navigation device. The measured 36 km/h could even easily be increased by using a larger number of accumulator cells or a motor with fewer turns. Because the mechanism of the Cross Tiger, however, is designed for about 30 km/h, it should not be excessively strained.

*Figure 73: Speed measurement with GPS navigation device*

Even after a long driving operation with frequent braking and acceleration phases, the motor temperature rises only slightly above 40°C. The included brushed motor is getting significantly warmer with a similar driving style even with weaker acceleration and a smaller top speed. This is also again due to the indicated efficiency of 89% of the LRP Eraser. It is by far superior to that of the brushed drive.

**Voltage curve**
While the trapezoidal voltage is absolutely necessary for the sensorless brushless motors to measure the phase zero crossing, this drive with Hall sensors could also be operated with a normal sinusoidal voltage, as described in Chapter 2. This could then also be implemented with a pulse width modulation. However, as shown in Figure 74, also here the trapezoidal voltage is used. The motor is operated there at full throttle.

The rotational speed is calculated with this two-pole inrunner to 1/0.0036 x 60 rpm = 16,666 rpm. The calculated maximum speed of 7.2 x 2800 rpm = 20,160 rpm is not reached, because the complete mechanics is moved along and therefore friction losses are caused.

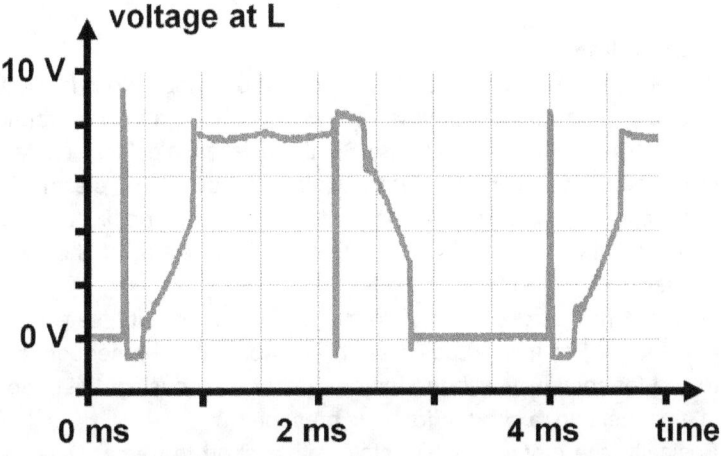

Figure 74: Also the LRP drive is controlled by a trapezoidal voltage

# 6. Sources of error

If the motor doesn't rotate or when it is already hot at idle running, if it makes noise or disturbs the reception, then the debugging begins. Basically, the error could be anywhere in the entire drive chain between battery and motor shaft. But it often turns out that it is in the motor itself or possibly in a defective power stage of the controller. On the one hand this is therefore calming, because not so many possible causes of error exist, because of the relatively simple motor structure. On the other hand, troubleshooting at the controller is much more difficult. You'd have to find the errors very quickly at the level of electronic components. This is not everyone's cup of tea. In the next sub-sections some typical causes of errors are described.

## 6.1 Short to frame, short-circuited coil, short circuit

**Short to frame**
A short to frame is present if the insulation of a copper coil is blank at one point. This usually happens when winding the wire around the sharp-edged stator sheets. So there is an electrical contact with them. Since the stator, which consists of electrically conductive iron, produces a contact to the housing, the ball bearing and the shaft, the entire motor lies on the voltage of the bare copper.
One can measure a short to frame with the continuity tester of a multimeter. For this purpose, all three cables can be sequentially tested for continuity with the motor shaft or the housing. If it beeps in a connection, a short to frame is present.
Basically, the motor can still rotate with a short to frame. Only the electrical equivalent circuit of the winding changes slightly, as a whole motor hangs on the coil. The impact on operating characteristics is usually not noticeable; the motor is still running smoothly. Thus, a single short to frame is often not even noticed.
It only becomes critical when multiple shorts to frame are present simultaneously. If a winding touches the stator at two or more

blank positions, the windings in between are short-circuited and the motor generates less power and torque with that coil. Another form of a short to frame happens when two different windings are bare and make a short circuit over the stator. Depending on the location of the shorts to frame, the result in both cases can be that the motor will no longer rotate or the controller notices the unbalanced winding ratios and turns off immediately. At best, they only have a bumpy run behavior. They can then become noticeable that the motor only starts in certain rotor positions.

As shorts to frame always occur in the innermost windings, which touch the stator, they can't usually be localized exactly. They can only be fixed if the corresponding winding is renewed completely.

**Short-circuited coil**

A short-circuited coil is either as described above, two ground shorts to frame of the same coil or a short circuit between two turns of a wire. The wire is then bare at two windings lying over each other, and all intermediate windings are shorted.

This case is more difficult to notice than the short to frame. The winding resistances of the three coils must therefore be measured (for the procedure see Section 3.2). If one of them differs strongly from the other two, a short-circuited coil exists. A short-circuited coil has similar effects as the short to frame, and the correction is the same as well.

**Short circuit**

A short circuit is the most serious form of all shorts. Here two of the three motor cables are temporarily or permanently connected to each other. For example, it can occur in an inadequately insulated connecting point between the controller and motor. Sometimes even bare cables at the housing inlet lead to a short circuit and simultaneously to a short to frame. A short circuit is detected by using a continuity tester. Otherwise you can measure it in the same way as the short-circuited coil. In contrast to this, the measurement result of the resistance between two of the three cables not only strongly differs from the others, but is also nearly zero.

A short circuit means that the current through the motor cable and through the corresponding MOSFET of the power stage will be very large. According to Ohm's law, I = U/R and with an assumed R = zero, the current would be infinite. Only the very small resistors RDS_ON of the power stage and the motor cables reduce it somewhat. Not all of the controllers available today are so-called short-circuit proof. Especially those which don't have a current measurement have no chance of switching off the power stage in time. Also, some controllers with current measurement react too slowly here. The result is that either the shunt or one of the transistors burn through and the controller will be damaged.

## 6.2    Bearing and shaft

**Bearing**
The bearing is the only connection between the shaft and the stator and thus the only mechanical wearing part. Most bearings are realized as ball bearings with steel balls. These bearings are often designed to be self-lubricating. This means that the lubricant is already placed in the manufacturing process and remains in the housing of the bearing during the whole operation time. Maintenance or lubrication is therefore no longer necessary. In extreme cases, special motors can rotate at 50,000 rpm or even higher. Then special ceramic bearings can also be used. In the case of motors with a small power, these can for example be journal bearings, while in motors with a higher power ball bearings with ceramic balls could be used. Other special bearings include magnetic bearings. These are then completely without wear. When used within the operating specifications of the manufacturer, the bearings should work long enough. The bearings in industrial applications must be designed for permanent use. Therefore they must have a very long life span. In model construction in contrast, only a few motors have a total operating time of more than 100 hours.
But there are aspects that fundamentally affect the bearing lifespan. They are:
- overly high motor power combined with incorrect warming
- unbalance of the load, especially the propeller
- unique mechanical shocks, for example caused by crashes

A defective bearing makes itself felt in that the shaft can no longer be rotated round by hand. We then have the impression that there is something like sand in the bearing or we feel a significant bearing play with cracking noises.

The bearings always sit with a press fit in the stator. To replace, you can beat them out carefully with a hammer and a center punch on a bench vise. Several manufacturers also offer these bearings as spare parts for their drives, even if they don't always like to see that the customer beats around with a hammer on their motors.

### Shaft

Under extreme stress and shock, for example crashes, the shaft can be affected. It is then bent and wobbles. Many model constructors have probably already tried various methods to bend such a shaft straight again, and not a few have failed with that. The following advice may not appear satisfactory: only a few model builders have the proper equipment to bend motor shafts straight again. Maybe the time is then better used for choosing a good new motor.

### 6.3  Defective power stage of the controller

The power stage of the controller is almost never defective by itself, but always by a faulty operation, as described in Section 6.1. In this case, the defect is usually diagnosed by the microprocessor. When trying to start the motor, no current flows in the corresponding power stage, and there is never a phase zero crossing. The defective MOSFET can usually be found easily. It often has visible damage and smells burnt too. Anyone who is versed in soldering can easily desolder the defective component and replace it with an identical new one. It usually takes a little power, since the transistor always sits on the thickest conductor, which leads the major motor current.

## Epilogue

Now that all the basics of brushless drives have been developed, now that all the advantages of this technology lie on the table, we might consider what the future will be.

Model construction is and always was a springboard for new technologies. Everything you can do in 'small' is always developed and produced much more cheaply and more efficiently than in 'big'. This also applies to the treated subject. However, nowadays 'brushless' characterizes our everyday lives, be it in CD and DVD drives, which must perform their service wear-free and reliably, or even with ordinary hand drilling machines or cordless screwdrivers. Industrial actuating drives are today also almost always 'brushless'. Electric cars will change our world sooner or later, because fossil fuels will be exhausted in the foreseeable future. Here too, model construction already anticipated the future some years ago; the electric motor has always been used in RC cars. Also for 'real' electric cars, the brushless drive is the standard drive at the beginning of the 21st Century; it has the name EC (electronic commutated) drive or synchronous drive. There are currently only very few similarly efficient drive alternatives which don't require fossil fuels. Already the first brushless motors are even being used for manned aviation.

More efficient energy storage devices than the LiPo accumulators have long been present in the minds of researchers or are already being tested in research laboratories. The model builder is assured that he can be one of the first to also use these in his models. When these new energy storage devices gain marketability in the foreseeable future, then the combination with the brushless power advances to the standard for everything that moves in our world through technology, whether autonomous or site-specific.

Of course new problems will arise. For example, sufficient and clean electric power sources must be provided. In addition, the new drives will also need new resources; with lithium or neodymium just two examples are mentioned here.

The technique will thus produce many new achievements. However, it should be the goal of any technician or engineer to use it so that it makes our world more pleasant. Brushless technology has the potential to do so!

## Literature

**Hanselman, Duane C:** Brushless Permanent Magnet Motor Design. Magna Physics Pub, 2006. ISBN 978-1-8818-5515-6

**Hilgers, Heinrich:** Selbstbau von Brushless-Außenläufer-Motoren praxisnah erklärt. Neckar- Verlag 2006. ISBN 978-3-7883-0683-0

**Hughes, Austin:** Electric motors and drives: fundamentals, types and applications. Elsevier/Newnes, Amsterdam 2006. ISBN : 0-7506-4718-3

**Irving, Gottlieb:** Electric Motors and Control Techniques. McGraw-Hill/TAB Electronics, 1994. ISBN 978-0-0702-4012-4

**Kafader, Urs:** Auslegung von hochpräzisen Kleinstantrieben. Verlag Maxon Academy, Sachseln 2006. ISBN 978-3-9520143-4-9

**Krishnan, Ramu:** Permanent Magnet Synchronous and Brushless DC Motor Drives: CRC Press, 2009. ISBN 978-0-8247-5384-9

**Passern, Ulrich:** Das LiPo-Buch. Verlag für Technik und Handwerk, Baden-Baden 2008. ISBN 978-3-88180-781-4

**Retzbach, Ludwig:** Deutscher Modellflieger Verband Brushless-Fibel II. Wellhausen & Marquardt Mediengesellschaft bR, Hamburg 2005.

www.ingramcontent.com/pod-product-compliance
Lightning Source LLC
Chambersburg PA
CBHW071210240526
45470CB00018B/1700